I would like to thank

all of my partners, who gave me the incentive that was so important in writing this book, and have been a great help in reading, correcting, formulation, etc. You are simply marvellous. Many thanks also to Wissi, who does not begrudge my success and makes me believe I am doing the right thing. I also hope this book will take my partners a stage further in their development.

"What you are
is God's gift to you,
and what you make of yourself
is your gift to God."

Danish proverb

E-Mail: info@gabi-steiner.de
www.gabisteiner.de

Distributed exclusively by:

Andreas Steiner e.K. Weinstadt
E-Mail: info@andysteiner.de
www.andysteiner.de

Translated by: Andrea Champell

Gabi Steiner

From Person to Person 1

Earning a Stable Income from Referral Marketing

ISBN-Nr. 9783945261026, 4. Edition August 2015

Contents

Foreword

"My name is Gabi Steiner. I was a single mother for eight years and was 41 years of age when I first met my partner Manfred. We were therefore extremely conscious of the value of time right from the word "go". Neither of us wanted to work until we were 65, just for the sake of possibly spending a few enjoyable years together. Our goal was no longer having to work by the time we were 50 – in other words, being able to choose when we want to work, how much we want to work and above all, with whom we want to work!

In 1999 I discovered a way of achieving this goal without investing money and without taking risks."

This book is my way of telling you about this opportunity. I would like to show you that you can achieve a great deal that you don't even dare to think about at present. I would like to encourage you to learn to dream again.

Various different people came up with the idea behind this book quite some time ago. "Leaders are readers" – particularly in our line of work, it is an invaluable advantage that everyone who is successful leaves behind so-called "tracks." Plenty of books and articles have been written by people who like to explain how they did it. Right from day one I have been enthusiastic about the possibility of studying the formulas of extremely successful people for no more than a few Euros.

Unfortunately there was a slight problem: all these books describe multilevel marketing, which means that they usually describe a mixture of direct selling and network marketing (team building), which has serious repercussions on working methods.

I still haven't been able to find a book about referral marketing that explains the type of marketing that our company practices. In other words, every book that is published requires additional explanation, such as: *"The information in this book is quite good, but you can forget pages so-and-so to soand-so. We don't practice direct selling, so it doesn't apply to us…."*

The endless need for explanation – I now realize that one of the most important aspects of our work is to make people understand the important difference between these two selling methods.

I am convinced that many of our partners still fail to understand precisely this point, and for this reason they cannot see the gold mine that is lying at their feet. This is also partially due to the preconception that most people have in their heads with regard to "this selling method."

My brother Andy had major problems with his company in the building industry and watched me becoming increasingly successful over a period of four years. When he was ready to talk to me about my method in July 2003, I first gave him a problem to solve. It was clear to me that he was extremely biased, and that I would have to motivate him with regard to this opportunity. I therefore asked him to solve the following problem first. I explained to him that he would have to open up a "new pigeon hole" for that which I had to say to him.

This task was such an important stimulus for Andy's decision that I have used it on frequent occasions since, and I would also like to show it to you.

Here are nine dots. Please try (before you turn the page, of course) to join all nine dots with just four straight lines – without removing the pen from the paper:

Andy couldn't solve the problem, of course – can you?

This is how it works:

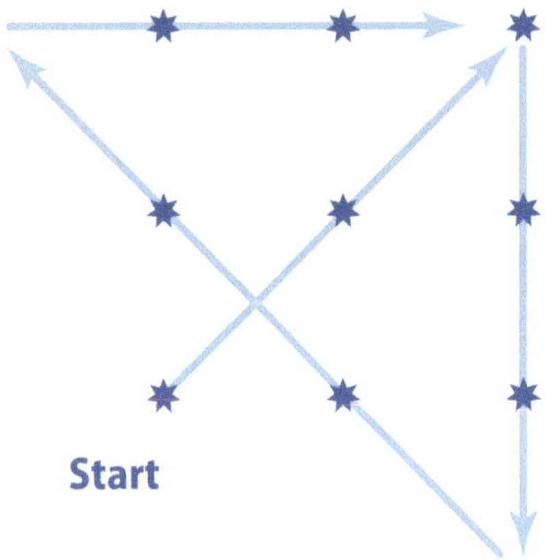

It was interesting, because he understood immediately what I was trying to say to him: "You have to think beyond the boundaries!" I gave him one of my favorite books and the following day he sent me an e-mail that I found deeply moving.

> "Have already read most of the green book. Sonja too! It is quite amusing when you think about yourself and discover that you only manage to join up the outer dots due to habit and laziness, and never reach the most important dot in the middle. You do this until you grind to a halt and then wonder why. It is time to change this and I hope that we can achieve it with your help!"

Can you image how I felt? I was motivated to write this book by this particular feeling, the need to provide explanation because of the existing preconceptions, the burning desire to use my experience to encourage one or two people to take the route to freedom and independence and a multitude of other reasons. Of course, I am not trying to say that you should ignore all the other books! One thing that I have always tried to do is remember a sentence from each book that was useful to me in one of my conversations, and may have provided the crucial argument with my conversation partner.

As a newcomer you will find many stories that may inform and inspire you when you start out. In accordance with the guiding networking principle of *"All power to the beginner"* I have decided to put everything that is important for decision-making in this book. At the same time the book is intended be a reference volume (a so-called starting aid) for helping our beginners to cope with their first year in this new, exciting world.

All the examples are real stories from people who have made their decision and are either on their way to achieving their goal or have already achieved it. I would like to thank all of my partners who gave me their stories and have therefore contributed the contents of the book.

Gabi Steiner

What is Referral Marketing?

A way of earning money? It always saddens me when I hear that someone has trivialized this opportunity into "earning money." Apart from prosperity, I see it more and more as an opportunity to acquire ideal values such as a considerable degree of freedom and independence. Real wealth means helping other people achieve success, making and maintaining friendships, making new acquaintances, learning about new traditions and customs, and above all having the luxury of "time" to devote to health, family, friends and hobbies.

The biggest challenge is to convince the person we are talking to that we actually have nothing to do with selling. Of course there are products which are sold. A respectable company always manufactures products and sells them. This is their whole purpose. But: WE do not sell. It is the company who sells. We do not collect money nor do we have anything to do with the flow of goods. Therefore, I would like to start by telling you a story that explains how I currently differentiate between traditional selling and Referral Marketing:

In July 2004, I wanted to take a 'holiday in Switzerland. Thanks to Lissy and Werner's family, (you'll hear more about them later) my group had expanded in Switzerland, and the Swiss team was excited about the fact that I was going to hold a couple of seminars there during my stay. The first seminar in Zurich was somewhat "terse," which wasn't helped by the fact that there was a woman sitting in the first row who had obviously made her mind up that she wasn't going to enjoy it even before I arrived....

Now, I am a passionate speaker, and I love to have people sitting in the audience who want to listen to my experiences. On the other hand, I have to admit that I am extremely "gut-instinct oriented" and sensitive, and this "vibration" was therefore extremely off-putting. (Only someone who knows me would have noticed this – but I found it extremely difficult because I had to think about every sentence. When I am in "full flow," on the other hand, the words just pour out).

After the break, the chair was empty and at the end of the event the sister of the lady in question came up to me and said, *"I don't know what to do with my sister – she thought that the seminar was just going to be about selling."* The nightmare rears its ugly head yet again!

I have learned that situations that appear to be unpleasant are often challenging or contain growth potential. I racked my brain for the entire week in Switzerland and tried to come up with a better way of explaining Referral Marketing so that everyone would clearly understand. And then something occurred to me....

On Friday evening, the seminar was in Lanquart, Switzerland, and I simply reorganized my programme and told the story about the woman in the first row, explaining how I felt.

What is Referral Marketing? It is quite simple, but before I explain it, you have to realize one thing: For each product that you buy in the shops, be it a book, the trousers that you are wearing or whatever, you pay the shop price. Let us call this 100 %. The question is, what percentage of the money actually goes to the manufacturer? I usually ask people to guess, and most of them come up with a figure of 20 to 40 %. Let's just split the difference and say that the manufacturer gets 30 %. However, at the same time, this means that 70 % has been lost in the distribution chain. This money is used for advertising and maintaining the distribution chain, for example. The rent for the shop has to be paid, irrespective of how much turnover has been made. This is why many self-employed people suffer from "fixed costs." Employees still have to be paid their salaries, even if sales have been somewhat down

Most people are extremely aware of this. In Switzerland, I gave them the following example on this particular day:

"Just imagine that you are on a street with three filling stations. One of them is called 'Ruedi Rüssel' (don't laugh, there really is a filling station with this name in Switzerland), the second one is 'Shell' and the third filling station is a special one. This filling station doesn't have a building, just a lonely pump, and if it rains, you get wet. There are no employees to come and serve you – it is self-service only.

However, it represents a unique opportunity: the money saved on employee salaries, service and rental or leasing costs (and this is quite a lot of money) is paid out to the people who recommend this filling station. In other words, if you spend 100 Francs on petrol at this filling station, you get back a certain amount at the end of the month for every person that you tell about this who subsequently refuels at this filling station. Let's assume in this case that you received 10 Francs for each recommendation. This means that when you refuel in the first month and tell your friend Anne about this special filling station and she refuels there as well, you receive 10 Francs. In the following month, let's say that you mention the filling station to your father Alfred, and Anne tells her cousin Bernard. Now three people (Anne, Alfred and Bernard) are refuelling as result of what you did. This means that you are now getting 30 Francs or Euros back."

The answer to my question of *"How many of you would refuel at this filling station? "* was almost 100 % unanimous in favour of my "special" filling station. Now let's take it a step further. I asked my Swiss attendees (who weren't quite so terse at this point) whether they could imagine recommending this filling station to one person every month. They all said yes.

When I came up with the following calculation, they were flabbergasted. Presumably, this happened at the same time as the breakdown of the preconception that most people have with regard to this matter.

"In the second month, four people refuel, including me. Like everyone else, I pay my 100 Francs for the petrol, but receive 30 Francs back (three people times 10 Francs). If everyone else recommends another 'refueller' per month, and this person refuels there, eight people are refuelling in the third month, and 16 in the fourth (at which point my own petrol is paid for and I still have something left over). In month five 32 people are refuelling, in month six there are 64, in month seven there are 128, in month eight there are 256, then 512, 1024, 2048 and in the month twelve no less than 4096 people are refuelling.

Four thousand ninety-six people are refuelling, even though I only recommended the filling station to how many people? Correct! Just 12 people. My friend Anne recommended the filling station to 11 people; her brother Bernard recommended it to 10 people in the example, etc. This is the power of multiplication. And this produces a total for which we really do have to open a new 'pigeon hole.'"

And now… the all-important question: "How many of you would seriously feel that you were selling petrol?" I wish you had been there in Switzerland. It was unbelievable how you could hear pennies dropping by the dozen. **That's it!**

Sometimes I hear the argument "***But something is actually being sold.***"

That's right! Of course petrol is being sold but not by those who recommended the petrol station!

People here in Switzerland had spotted the opportunity to achieve a small or large income by recommending this filling station. And this is the reason why Referral Marketing companies exist. Simply because there are now enough reasons to look for new possibilities. Our old methods no longer work. Just think about your jobs and your pensions, and what about your health?

Referral Marketing is the solution to many problems. I often ask myself why many people fail to see this. Perhaps it is because of the human trait of having to find something wrong before we can find it right.

It was nice to see them scratching their heads when I asked, *"How would you tell your friend about this matter? Would you emphasize the quality of the petrol and tell them how they can refuel with excellent petrol with a great octane number and a golden-yellow color? Above all – would it occur to you to say, 'I haven't got the time?"* You are probably laughing now, because you are now aware of how amusing that would be.

Discerning readers will have found two catches in the example: the first is that you cannot pay 12 x 10 %. This is clear. Most companies pay out a smaller percentage the "deeper" it goes. The total is usually about 60 %.

The second catch is it doesn't work with petrol. This is probably because of the sign that I recently saw at a filling station that said, *"We are not petrol distributors, but tax collectors."* It doesn't matter – what matters is that you understand the concept. And I hope that it is interesting enough that you will join me in considering a product with which it would work.

It should be clear to everyone that **turnover** has to be generated in order to pay out commission or bonuses. My Swiss people were now ready to start thinking about a product other than petrol that would be suitable for this marketing method.

I asked the question *"What characteristics does a product need in order to be suitable for this method?"* Let's think about this together. It must be a **consumable** product, of course. A vacuum cleaner is not ideal, because it doesn't disappear in a puff of smoke every month. It therefore has to be something that is "used up" every month. This is crystal clear, because otherwise you cannot have a **residual** income

Of course, "residual" does not mean that it starts to rain money without having to do something for it. Like any other business, you can only have a steady residual income if you have done something to earn it beforehand. In referral marketing, this means helping your partners to understand the business until they are in a position to continue for themselves. Essentially it is a case of providing people with support and helping them to build up their business. The more successful you are in doing this, the more independent your income will become from your efforts. After all, this is the reason why most people start to do referral marketing in the first place.

It is also important for the product to be suitable for **everyone**. Horse fodder may well be consumed, but how many people own a horse?

And now, the most important point; it must be something that is **important**, something that we really need, that is in vogue and in an area with growth potential and a future. What are the options? This resulted in an interesting discussion with a unanimous conclusion: Only one area comes into question, and that is wellness, fitness, health and anti-aging.

And since my company operates in precisely this so-called "trend market" or "growth market" with natural products, I would like put forward the thesis that **everyone** is interested in our products. You should have heard the protests… (Which was actually my intention).

In spite of this, I stand by my opinion. I still maintain that everyone is interested in "Living Healthier for Longer," like the title of a book written by Anne Simons that describes OPC, which one of our main products. The problem lies elsewhere. The problem is that only a few people already know or believe that there is a definite connection between the diseases of modern society, our aging process and certain foodstuffs. Why is this the case? Scientist Max Planck explained the situation in a nutshell:

"Scientific truth does not assert itself by convincing its opponents, but by the fact that its opponents gradually die out and the next generation grows up with the new ideas right from the start."

It is a pity about the large number of people who grow up, live in the "interim period," and are unfortunately unable to experience this.

But what if there really is something in it? What if the countless nutrient and antioxidant studies are correct, and you do not react to this stimulus or even check it out, or do not even bother to obtain additional information? With your hand on your heart, wouldn't it make sense to follow up such an important tip? Can we afford not to?

I told the Swiss people a simple story that almost everybody knows, *"What happens to an apple when I cut it open?"* Correct. The surface turns brown. *"Why?"* Many people knew the answer – it is because of the oxygen, the free radicals and oxidation process.

The same process causes iron to rust. *"What can we do to stop this 'browning,'"* I continued. Most homemakers know the answer: put lemon juice on it. *"And why lemon juice? What's in it?"* Of course – vitamin C. And this vitamin C prevents the oxidation, "aging" or "rusting" of the apple for about another four hours, because it is an important antioxidant.

Imagine that the latest research on this subject was correct and antioxidants (also known as radical neutralizers) could have the same effect on our bodies.

At this point, I always give my conversation partners a piece of advice: Just fill up with 'Premium' instead of 'Regular' for a month and see what your body makes of it.

What if you could see the proof of the literature on your own body – could you keep it to yourself and **not** pass the information on to your friends and relatives?

I often think about the history books that will be written in the year 2050. I imagine that they will say: *"At the beginning of the 21st century, people had already looked into the effect of antioxidants and therefore found a solution to the serious problems caused by the diseases of modern society. Inexplicably, a mixture of ignorance, complacency and inflexibility meant that it was decades before this knowledge became generally known and utilized...."*

Sounds logical, doesn't it? I presume that you realized that it is worth getting a deeper understanding of this topic, and that you would also want to try it all out if it didn't sound like a pyramid or snowball system....

Snowball system or pyramid scheme?

Another nightmare in our industry that causes every new starter's blood to freeze in their veins is the question *"Is it something like a snowball system?"*

This question is extremely important, and the uncertainty surrounding it costs many potential Networkers their livelihoods. For this reason, I would like to deal with it in detail right at the beginning of this chapter. Professor Zacharias, who lectures on the subject of Network Marketing at Worms University, gave important stimulus for this in his pamphlet "Die Wachstumsbranche der Zukunft" ("The growth industry of the future").

This is certainly not a chance remark, since there have already been several cases of companies that were not entirely trustworthy. These have been eradicated by the law, which has been watching them like a hawk ever since. According to Professor Zacharias, the typical features of a snowball sytem are:

1. Commission is earned by recruiting new contract partners, making the actual selling a secondary issue.

Payment in Network Marketing, on the other hand, is dependent on turnover.

2. In a snowball system or pyramid scheme the products are purchased from the next level up, and sold from level to level with a surcharge (this means that Anne pays 10 Euros, for example, and sells it to Bernard for 12 Euros, who sells it to Christine for 13 Euros etc.).

The difference between a snowball system and Network Marketing is that in Network Marketing the products are purchased directly from the manufacturer – and at the same price at every level of the hierarchy!

There is only one thing to say about pyramid selling: Every company and institution has a pyramid shape. Every company has a CEO at the top, below him are the directors and under them the managers of the different departments etc. etc. The same applies to the British Government, which has the Prime Minister at the top, then the ministers, the members of parliament below them and then the broad mass of the population at the bottom. So let´s get away from the idea that a pyramid is something that is inherently disreputable.

So, how can you tell whether this way of earning income is a legal business or an illegal pyramid scheme? Certainly **not** from the structure of the sales organization of the company that has been described, since this has a pyramid shape.

The WFDSA (World Federation of Direct Selling Associations), which was established in 1978, currently represents 50 national Direct Selling Associations (DSAs) at global level. This global association and all the national DSAs have always recognized the need for ethically proper business behaviour, and therefore developed a global behaviour code for the industry. The prerequisite for a global DSA is that the company complies with this code.

We can assume that companies who are members of a national DSA and sell their products via Network Marketing are not illegal pyramid systems. My company is a member of the DSA and was awarded the "Best New Business 1998" award in England when it was established.

What are the criteria for a legally operating company?

Products must flow!

It is very simple to distinguish illegal pyramid systems from Network Marketing. If the product flow from the company to the end consumer is horizontal and the company also supplies products to its associates under the same conditions it is classic, legal Network Marketing. The money also flows horizontally back from the end consumer to the company. In this case, WHEN the associate starts in the business and how many levels are between him and the company is irrelevant.

The other day, Michael Strachowitz, a renowned network trainer, explained a pyramid scheme in a way that made me chuckle. At the same time, however, it left me thinking.

"We are talking about a pyramid scheme, when the income of the members who are already in the system is generated by the enctrance fee of new members, the system, however, breaks down as soon as there are no more members coming into the system."

...why does our pension system spring to mind right now!

I hope that I have dispelled any fears that you may have had and that you can now carefully follow me as I tell you more about my story.

The importance of the story

"Network Marketing is the business of telling stories and relating personal highs and lows...."

"The power that the stories possess is a golden key that opens the door to people's hearts."

And:

"The right story at the right time awakes people from the narcosis of the mind, frees them from the slavery of old ideas and overcomes logical thinking. For one moment, all sensibility is forgotten; emotions get the upper hand and previous convictions are occasionally questioned. We find ourselves on a plane at which we can think beyond that which is rational."

These quotations from the book "Dream Teams," describe a truth that I have long since been aware of in a theoretical sense. But I must confess that I have only known for a few years **how** important this point is, and the **extent** to which this knowledge can be transferred to our business. Nowadays I consider my own personal story to be the main issue.

The primary and most burning question that is asked by every newcomer is *"How do I talk to people in my environment?"* To be honest, it doesn't matter as long as we do talk to people.

We only know one thing for certain from experience: Someone with passion and enthusiasm but no knowledge will have a better start than someone who can mechanically recite all the facts and figures to perfection. We have homemakers with six children and no previous knowledge who can uproot trees with their enthusiasm. On the other hand, we have qualified engineers and sales professionals who "already know it all," making them less prepared to learn, which leaves them unsuccessful. On several occasions, I have found that people who are extremely successful in his or her normal professional life have been unsuccessful in referral marketing. The only reason for this is that their pride does not allow them to accept such simple and straightforward things from a homemaker....

The prime directive in referral marketing is therefore:

Never judge anyone based on his or her previous success or knowledge. Never decide whether someone is suitable for the business.

We have a saying that goes:

"The person who you think is a "doer" will be a non-starter, and the person who you wouldn't have believed it of will astound you."

Basically there are many ways of talking to people. At the end, it is the **quota** that matters. But one thing I can tell you for sure:

- The more I speak with people the more they ask me what I am doing.
- The closer I am with someone (the "warmer" the contact), the greater the credibility and the interest in what I am doing.
- The better I know their **"WHY"** the more likely it is that the business will be a solution for them and therefore the higher the quota.
- The better I am at "active listening", the higher the quota.

In this book, I have decided to tell you about the things that have the best prospects of success. In my opinion, referral marketing is a person-to-person business, and I particularly like it because it gives **everyone** the chance to be successful, regardless of age, sex, profession, or origin. I therefore prefer to recommend and teach methods that can be **achieved by anyone** and therefore be duplicated. However, this certainly doesn't mean that other methods do not work. Only one thing is certain: Even if you are one of the few people who does not have a problem giving presentations in front of large numbers of people, you should consider one thing: 80 % of your group will consist of people who cannot do it.

Another thing of which I am certain is the less we know about the people to whom we are speaking, the more conversations with people we will have to conduct. That is the reason why I like to talk to people who I know. That is called the "warm" market. Of course, I can get to know any person – I always like to say *"any 'cold' relationship can be turned into a 'warm' one."*

Partners who are new to the business are often so enthusiastic that they talk too much or say things that their conversation partner is not interested in. When this happens, there is a risk that our conversation partner feels steam-rolled by the flood of information and goes into defensive mode.

The most effective and harmless way of awakening interest is to tell personal stories. If you tell your story in an authentic and interesting way, your conversation partner will seldom be able to resist becoming curious and asking you questions.

We tell people why we are in the business, what persuaded us, how we started and about the opportunities that we envisage of putting our destiny in our own hands. Bridges between people can only be built using personal emotions. Jörg Löhr, a well-known professional trainer, said something that I have never forgotten: *"Our era is dominated by communication and emotions. Machines have already replaced our muscles, and computers have replaced our brains. The only thing that is exclusive to human beings and makes us unique are our emotions."*

I am a shy sort of person and therefore have to tell my story and use it as a bridge. (You will already have read my short story at the beginning of this book). With my story, I am making an indirect offer and am therefore planting a seed. My business partner therefore has the opportunity to be responsive to it or not – and believe me, someone who is on the lookout for change will ask me questions.

That is another advantage, by the way – **THEY** are then asking me questions, and **I** am not offering them anything. See the difference?

The story of the head doctor in a city hospital who has a lot of money but no time to enjoy his life would go like this, *"You know my situation – money isn't my problem, but I have been thinking about whether I really want to carry on working in the hospital for the rest of my life. My marriage is on the rocks and my children hardly know me. I have now discovered a way of making myself financially independent without having to give up my profession, and without having to forego my security."* His conversation partner may now have become curious and says to the head doctor *"That sounds interesting, tell me more about it."*

If we have understood that we can make referral marketing work better by telling our story, it becomes easy to talk to people. And if my conversation partner is interested, he or she will ask me questions. This method (if telling stories can even be called a method) is absolutely stress-free and unrestricted. We can talk to someone in a relaxed and unforced way. Products are not part of the story at this point.

Richard Poe explains the reason for this in his book "Wave 4 : Network Marketing in the 21st Century":

"Every salesman is a storyteller. In most cases, salesmen tell nuts-and-bolts stories about the uses and benefits of the products or services they sell. Network marketers tell a different type of tale. They talk about themselves, their lives, and their goals, dreams, and aspirations."

Do you know the difference between an inexperienced and an experienced Networker?

"The experienced Networker knows more stories!"

This is true. This person can tell a suitable story from his or her repertoire at the drop of a hat. This book is the same – it contains many stories with the marvellous ability of making you remember them long after pure theory has been forgotten.

When I am carrying out starter training with a new partner, the partner already has at least two stories to tell: Mine, which he must use until he has his own income, and his own story. As a sponsor, one of my first duties is for us to "put together their story." Of course, this does not mean inventing one. It is more a question of finding out the "**WHY**" of my partner, or their "primary motivation factor," as it is referred to by Allan Pease, the author of the best seller "Why men don't listen and women can't read maps." This is the "burning issue" as far as we Swabians are concerned. At the end of the day, it is a question of discovering what it is that is so important to my conversation partner that it spurs them into action.

*Starter training is the initial training that I carry out with a new partner who has made the decision. It is a case of explaining the order and filling it in, if this has not already been done, explaining the initial steps and simply answering all the questions that currently exist. An up-to-date starter training specification can also be found at: www.mitgliederbereich.com.

My detailed story

When I first encountered this industry in August 1993, I immediately spotted the opportunity that it presented.

There was no doubt in my mind whatsoever that I would be successful using this system. I detected immediately that it was dependent upon my input and **nothing else**, and I was prepared to do it. In other words, I was prepared to pay the price upfront.

I realized that I would have to work hard for a few years in order to enjoy a stable income, which is every Networker's goal. At the time, I was a single mother to my eight-year-old son Tim and was working part-time for a wholesaler of technical equipment. Because of the time that I needed to look after my son, it was almost impossible to build up a career. Nor did my financial situation give much hope for any kind of highlight in my life.

After just six months of working with my partner company as a sideline, I was able to give up my main job, which was a major relief, even though I had worked for the company for sixteen years with considerable commitment. It sounds crazy, but I enjoyed doing my job for all those years. Working Sundays or overtime was no problem for me when it was necessary.

In spite of this, the final days that I had to work in the office because of my three-month notice period were torture. I had heard this from many people who have been in the same situation. As soon as you "taste" freedom, as we say in Swabia, and discover how much fun you can have in referral marketing, "regulated" work becomes difficult to perform.

The diet/food supplement products made by my former company suited me. Since 1972 I have had a corn mill, (the owners of corn mills usually wore Jesus sandals at the time) and the connection between nutrition and health has been an important topic to me for years. The connection between nutrients and the aging process has since become my main hobby (responding to the need...).

I enjoyed my work, and because of my knowledge I quickly became a product speaker and spent many weekends at seminars. I now think that a great deal of valuable time for bringing up my son was irretrievably lost because of this. This is one of the few things in my life that I would like to have changed, because I now put a great deal of value on making business integrate harmoniously into family life.

In 1996, something happened that changed my life completely. I met my partner Manfred (who I call "Wissi"). To be precise, I pretty well threw all of the resolutions and opinions that I had about relationships and men out of the window and simply loved him.

Then the problems started. The first problem arose because his main working hours were during the week, whereas my main working hours were at the weekend because of my regular two-day seminars. The next challenge cropped up because Manfred had booked a three-week holiday in South Africa before we met. For me this was **much** too long, because I had invested a considerable amount of money in products from my company in order to get my commission from three levels. And how was I supposed to sell volumes like this if we were in South Africa for three weeks? On the other hand, the thought of celebrating our first Christmas 8,000 kilometers apart was unbearable. This was the first time that I had doubts about what I was doing.

At this time, I realized that it may have been a Networking company, but the main onus was on direct selling because of the compensation plan. And suddenly it was there – the realization that I didn't have any stable income.

I asked myself "What happens if I become ill? Or simply don't want to work any more?"

In spite of this, I continued with it for two long years until I met Don Failla by chance at a seminar.

I was at this seminar because all of our "tools" such as inserts, flyers etc. were no longer working effectively and many of my advisers were having problems selling their qualification volumes (= quantity of products that you have to sell to receive supplies for building up the group). No volume – no cheque. I wanted to give my group a new tool that would give them new impetus. My team mainly consisted of young mothers with children, and I think with regret about the early mornings that they spent posting flyers through letter boxes before their children were even awake. Summer' is not too bad, but can you imagine how it must have been in the winter? Don and Nancy Failla are extremely well-known Network trainers, and what they said at the seminar made a considerable impression upon me.

"Real referral marketing has nothing to do with selling. Your aim is only that a considerable number of people use their own product. You have a good product. Then you look for five friends with whom you wish to be successful and help them to talk to their friends. This means that you will never have to speak to strangers again."

Don's words *"Anyone can meet a stranger if he is introduced to him by a friend"* are still music to my ears. Everybody knows at least 100 people of whom at least five are certain that they want to make a change in their lives.

Earning money together **with** my friends, not **from** them. That is a big difference.

I was still enthusiastic about the simple Networking idea and immediately ordered 1000 copies of Don's book entitled "Ihre Zukunft" ("The 45 Second Presentation That Will Change Your Life"). To never have to use inserts again or talk to strangers, and never have to push flyers through letter boxes again. I immediately called my team together and explained to them the simple way in which we could now become successful. **NO MORE PUSHING**

PRODUCTS!!!!!, each one of us would have to sponsor just five friends and speak to their friends!!!

We all set to work in high spirits, but within a few weeks we discovered that nothing whatsoever was working. Why was this?

Because of the qualification volumes, we had to sell to earn our commission. And this was the first time that I really understood the difference between direct selling and Referral Marketing!!! And I didn't want to "push" any more. I wanted to be a teacher and show other people how they could really achieve their goal.

At this time, I read all the literature that I could lay my hands on. The following was written in a book by Edward Ludbrook:

> **"You must ask yourself the question of whether you can still earn money when you are no longer working. If you have to reply to this question with 'no' you are caught in the trap, like 99 % of all people."**

And there I was! I couldn't see a solution. At this time, I did not know of any company whose qualification volume could be covered by my own needs. Today I know that opportunities always come when you are open to them and ready for them.

My opportunity came in April 1999. I discovered an advertisement with a photograph of Don and Nancy Failla. Of course, it immediately clicked that if Don Failla was advertising for a company, it must be a company that fits in with his book. This mademe extremely curious, and I did some more research into it.

I was impressed by the fact that it was a company with roots going back to 1936. I was 44 years of age at the time and the fact that it was an established family company gave me a lot of confidence when I was making my decision. The product range, which did not use any chemicals, was another plus, and the bonus plan fitted in with my goal of really becoming financially independent. After six years in this field I realized that this marketing plan could be achieved without start-up costs and minimum purchase volumes for anyone, and could therefore be **duplicated**. I also realized that it was the kind of real **Referral Marketing** which I had dreamed of.

By the way my goal was set fairly low back then. As Don Failla put it at the time *"Imagine that the instalments on your house and your car are paid and that you would earn 5,000 Marks per month, regardless of whether you got out of bed or not."*

In the beginning this was my goal and I would have settled for an enjoyable life with Wissi. After all my previous experiences this opportunity was ideal to achieve my goal. It was a gift from heaven, like winning the lottery jack-pot being handed over to me on a silver platter!

The talking bird

Bill Evans, one of the company founders, once told a wonderful story: *"A man saw a bird that could speak 400 words and could sing arias in two languages. He was so fascinated by the bird and because his mother had a major birthday coming up, he decided to give her the bird as a present. He asked about the price, which was 50,000 dollars. This was a lot of money, but was worth it for a bird that could speak 400 words and sing arias in two languages. He had the bird carefully packed and sent to his mother. After several days, he called his mother and asked her what she though of the bird, and she said, 'It tasted lovely.'"*

How often do our new partners forget to say that they have a bird that can speak 400 words and sing arias in two languages?

Dreams and goals

In one of my favorite books "Owning yourself" by Paula Pritchard, it is written:

> **"Your career in Network Marketing should start with discovering your dreams and end with the fulfilment of these dreams."**

There is a lot of truth in this. Awakening dreams and formulating a goal from them is one of a Networker's most rewarding characteristics.

After ten years in the industry, I now know that careers are often nipped in the bud because many people have forgotten how to dream. Our dreams often end up in the smallest pigeonholes, are forgotten, turn to dust – perhaps as a result of frustrating experiences or just the daily drudge.

"There are no goals without dreams."

This saying is also extremely important.

In this chapter, I would like to give you the courage to resurrect your dreams and have confidence in yourself and your capabilities. And give yourself a real goal. I would like to show you a way in which to make work fun and, if you do it correctly, you cannot even call it work. A way in which nobody asks what training you have, or where you come from. Nobody filters you out because you are of the wrong age or sex.

Based on many practical stories, I will show you how to achieve the same thing that many others have achieved with me: Starting the day with a smile, being able to allocate your time as you wish, not having to worry about money and being surrounded by people who are positive and openminded.

I am fascinated by the idea of being able to build up financial independence together with the people who are important to me.

The fact that it works does not have to be proven any more. I've already done it...!

I started doing referral marketing in 1999 without investing money and without fixed costs. Nowadays I spend part of my time on my favorite island, Majorca, and could have stopped working a long time ago. (You will find some pictures and impressions of the island at www.gabisteiner.de). My monthly income was already far in excess of my previous annual income eight years ago....

As I said, it no longer needs to be proven that the system works. It is now more of a question of how it works, and above all how it will work for you.

This is another reason for writing this book. There is still plenty of room for those of you who have resurrected your dreams and turned them into a goal.

Too good to be true? Not at all. Because I also intend to show you that success and financial independence also have something to do with work, or to put it another way, with commitment, I will honestly tell you what is needed to get there.

I will show you that your own activities and work and particularly patience and endurance are required when you set off on the road to a stable income, and that there are hurdles to jump over and obstacles to overcome. But it's worth it.

The question often crops up of what I earned in the initial months. This is difficult to say, because I was and **still am** being paid for the work that I did in the initial months.... For example, the names Lissy, Erika and Matilde can be found on my second statement (Royalty & Infinity Bonus Report) from May 1999. Erika is now in the "Diamond" stage in Lissy's team, and Werner and Lissy currently have an estimated annual turnover of 10 million Euros. The turnover of Matilde's group is close to 3 million Euros. Can you imagine what this has done for my income over the last five years? And even better, what do you think my income **will be** over the next few years?

What an amazing opportunity! Can you imagine how many people miss a great opportunity because of something trivial?

There is only one difference, if 100 people in the free market economy apply for a job, perhaps five of them have the required training or qualifications. The others are either too old, too fat, too uneducated or too... whatever. They are **filtered** out....

If 100 people start out in referral marketing, they all have the same opportunity. In this case, they can filter themselves out:

- "Well, it's nothing but selling..." – we're already down to 99!
- "I don't have the time..." – 98!
- "My husband won't let me..." – 97!
- "It's a snowball system..." – 96!
- "I can't do it...." – 95!
- etc. etc. etc ...

At this point, I would like to tell you about a phenomenon about which I have been racking my brains for years. And I still haven't found the answer. If someone wants to be a mechanic or a nurse, for example, he or she decides to go through training at a point in time when it is clear that he or she does not know anything yet. They only know that they will now have to learn for three years to acquire all of the skills.

I have never known students to ask a teacher to give them all the necessary information **before they start** so that they can decide whether they want to learn the profession or not. Can this person repair cars or take blood during the first month? Of course not – they know and accept this and have **no** problems with it whatsoever, because that is simply perfectly normal. After all, a doctor has to study for five years before he can start practising. And if you ask the doctor when he started to work, it will always be the point at which he completed his studies. And he doesn't earn **anything** whilst he is studying. And afterwards? (OK, things were different in those days…).

It never ceases to amaze me that on the first day, at a point in time when they do not have any information or training, a person can be certain that they **cannot** do it. Do you know what I mean?

A few years ago I heard a saying that contains a lot of truth:

> **"Network Marketing is like driving down a motorway in the fog. You drive for 100 metres, and then you see the next section."**

We also need a training period. This also includes trying out the products and getting to know the industry. Dirk Jakob, who is one of our favorite personality trainers, advises everyone to set up a small library of at least ten "tools." Tools are CD's, books, cassettes, videos and brochures about the industry. On the one hand, this is important for our own security and our gut instinct, and on the other hand, it gives us tools that can be passed on. I know from experience that the team grows more quickly if more tools are deployed. There is no business that depends so much on our own attitudes and convictions than Referral Marketing.

> **"Regardless of whether someone believes that they can do it or not – they are always right."**

I also like to compare it with riding a bike. When we sit on a bicycle for the first time, we know how to use the pedals and in theory we also know how to steer – but can we ride it?

No! What is missing…? Correct! Balance! But once we have it – and this occurs in a fraction of a second – no one can take it away from us again. It is exactly the same in Referral Marketing, and I love the moments when I can share the experience of one of my protégés finding their balance and suddenly discovering that it is no trouble whatsoever. Conversations are suddenly fun and you talk because you are enjoying it and not because you want to acquire new partners.

I now know that there are rules and regulation that undoubtedly lead to success if you stick to them. And I am therefore certain that you can also achieve what I have.

"Apart from all of the financial benefits, the greatest pleasure that a real Networker can probably derive is experiencing how a partner in whom you have had faith has changed from being a timid, somewhat fearful person who had probably enjoyed an average amount of success in the past into a radiant, much sought-after personality who is in control of their own life, becomes an achiever and is therefore a role model and incentive to many others."

I read that quotation years ago from well-known Network Trainer, Michael Strachowitz, and I often think about it when I am standing on the stage at major events with goose bumps and realize how many people have blossomed. One example, who is now famous throughout Europe for her story, is Erika (remember my second statement?), a homemaker from the Swabian forest and the mother of five children. Once her children became independent, she wanted to do something for **herself**. She achieved Diamond status after three years, and when she stands on the stage and tells her story, you can feel how encouraging it is. This is also an extremely important reason why new partners should be able to see the "bigger picture" at an event or training course as soon as possible.

When they achieve a certain volume, our partners receive an invitation from our partner company in the USA. This occurred at an extremely early stage in Erika's case, and she was over the moon about it, of course. The neighbors in her small village soon noticed that travel preparations were being made, and when a neighbor asked *"Are you going to America on holiday?"* Erika proudly replied, *"No, it is a business trip."* Do you think that her friends and neighbors would have believed that she was capable of this?

Presumably there was a time when she would not have believed it herself! But she was always prepared to learn, and she had an excellent sponsor in Lissy.

"Networkers who can take on real leadership also firmly believe in the basic potential that each of their partners have within them. The partners themselves feel that someone believes in them, perhaps for the first time in their lives, and have so much more due to this fact alone."

And this belief is extremely important, because envy is not unusual, particularly in Germany. The saying *"Pity is something that you are given, but envy is something that you have to earn"* hits the nail on the head. I now know that to someone who has said "**No,**" success in our industry shows that he was wrong... and not everyone is happy about this. This is something else that you will have to learn to live with when you set off on your journey. I just wanted you to know this.

A goal is a dream with a deadline!

Before we start off in Referral Marketing, we need a goal. The goal must be something specific. We must be motivated by it. Only then will we be prepared for what we have to do. And withstand the knock backs and the envy that are bound to accompany it. This is simply my experience. Everyone would like more money, but only those of you who have a goal are prepared to set themselves in motion to achieve it. Big targets cannot just be achieved by recommending products, but only by setting up a stable organization. Managers are needed to do this. For this reason, we look for people who have goals.

If you advertise with products, you will get customers. If you advertise with a business opportunity, you will get business people. If you advertise with a vision, you will get managers.

Who is most suitable? Anyone who wants something that they do not yet have. And the more they want, the more suitable they are.

There are many books describing the importance of goals. I recently got my hands on one of them "Successful Goal-Setting" by John Church. I am so fascinated by the clarity of the examples that I have used them often since and would like to share some of them with you. John Church used to ask his seminar participants how many blue cars they had seen on the way to the seminar. Of course, no one knew the answer.

"OK, but what if I give you 10 Euros for the license plate of every blue car that you see on the way home." Would we see blue cars **now**? Of course, we would. That is the reason why clearly defined goals are so important. The goal notifies your brain about the opportunities that it needs to spot (blue cars). **If you don't have a clearly defined target**, your brain will have nothing to focus on. You will just meander through life and pass up on millions of opportunities that you could have used to achieve just about anything.

That is phenomenal. Something else that also baffled me is his question, *"Can you not achieve a goal?"* Can you? Think about it!

No, you can't unless you die beforehand.... You can change your goal if need be. Let's suppose that you goal is to lose weight. What you plan to do is change your eating habits and walk for an hour every morning. The next morning you get up and discover that your trainers are filthy, your tracksuit is in the wash, it looks like rain and you have so much to do that it's **difficult** today. Lack of willpower wins the day and you don't get your exercise. But you'll definitely do it tomorrow....

How often does something come along that seems more important and whoops, we have suddenly changed our goal? (Believe me. I am only talking about things that I have done myself...).

There is no such thing as failure, just success that didn't happen soon enough.

When you are striving to achieve a goal, it is extremely helpful if you can simulate the feeling that you will have when you achieve your goal. I found a good example in another book. It described a funny story that contains a great deal of truth:

Egon visits a car showroom and is standing in a trance-like daze in front of his dream car. The salesperson approaches him and invites Egon to test drive the car… Now there are two possibilities: If he refuses (no, no, perhaps not, I might want the car and I can't afford it), then the salesperson knows that Egon is a lost cause. But what if Egon thinks, 'Well, I can just sit in it, can't I? Just for a little while?' Then he sits in the car – the soft seat snuggles up against his rear end; he breathes the heady aroma of the brand-new leather upholstery. He starts the engine… Eight cylinders start to purr… he imagines what it would be like to drive up in this car at home. How his neighbour, who said 'no' to Egon's business opportunity right at the beginning is standing at the door – and how his chin drops… Suddenly possible solutions come to him. His brain works feverishly and churns out various possible solutions, 'I could give notice on the building society account, get Auntie Annie to pay out my inheritance, wash cars on Saturdays' etc."

Vision board

A good way of giving our brains specific instructions is to create a vision board. This is something that I like doing – particularly with new partners – and we always have great fun doing it. "Dream Building" – the awakening of dreams is like "listening" to an extremely well paid capability in Referral Marketing. The sooner our interested party sees Referral Marketing as a way of fulfilling their goal, the more likely they are to accept our offer. And as I have already said, I do not accept earning money as a goal – there are always things that you can do with it – and they are seldom material things.

I would now like to tell you my story from 1995, when I was still a single mother and my son was ten years of age. Four years prior to this in 1991, just before the big drop in property prices, I bought an apartment. My darkest year was 1995. Due to certain circumstances and a "smear campaign" against my former company, my income dropped to about 2,500 Marks per month within a period of six months. This is not an inconsiderable sum, but tragic when you have fixed costs of 6,000 Marks. Without a partner whose shoulder I could have cried on, and with the responsibility for a child who wasn't always easy to deal with, leasing for the car and the instalments for the apartment and negligible income, things were looking bleak. I could no longer work. Some mornings I simply didn't want to open my eyes. I didn't want to think.

At this time, one of my partners came to me and said, *"Either you do something to get us out of this mess, or I'm packing in as well."*

Since she was a member of my last two remaining active groups, her threat was a severe shock, and shook me up. I really couldn't afford to let that happen. She suggested going to a seminar held by a man who was achieving good growth rates in the same company in spite of the general downturn. His seminar was in Augsburg and cost 400 Marks.

"Are you crazy?," I said to Monika, *"I can't afford to pay 400 Marks to go to a seminar."* She then said something that was extremely profound that I still quote today. She simply said:

"You can't afford not to go to the seminar."

I went. And do you know what this man made us do? Cut pictures out and stick them onto a piece of cardboard. I was horrified – I scrape together my last bit of cash and drive to Augsburg, only to cut out pictures for a vision board. Good – I was no longer annoyed. I realized how important it was, and the two days were extremely decisive for me. He said a few things that touched me deeply and gave us an important formula that I digested at the time and still come out with at every opportunity.

Thinking + Acting = Success!

Let this formula run over your tongue. What does it mean? You can work like a Trojan if you have the attitude of *"it's not going to work anyway,"* and you will be correct. Or you can sit in the corner and think as positively as all the esoteric people in the world, but it's no good to you if you don't act. Strangely, I then started to work again, and four months later, my cheque was back up to the previous amount. What had changed? I must admit, the situation was no better. We had to do a lot of explaining and work harder to achieve the same success. We needed twice the quota. It was a slog. But my cheque continued to increase. We had hit rock bottom and learned something extremely important from it:

It is not the circumstances – they are the same for everyone. Nor is it the economy. Nor is it politics (although this is extremely useful as a scapegoat, of course). Nor is it the weather, nor fate and certainly not our partners. It is all to do with our attitude.

The goal collage that I created in those days has now been achieved (OK, I admit I still have to work on my BMI (Body Mass Index), but I am currently in the process). In 1996, I met Wissi. In 1999 I started my current partner company. In 2000 we travelled to the Maldives, in 2001 we bought the house at Ebnisee, and in 2002 I bought the small, flat, tellurium silver car on the advice of my tax adviser. In 2003 a dream was realized that wasn't even on my collage – a second home in Majorca. Now I urgently needed to create a new collage. It was created in 2004 – and I am carefully thinking about what to stick on it. The human brain is unbelievable. Let it work for you.... Imagine how you will feel when you achieve your goal. You will then develop a burning desire to achieve so much that you will be in a position to change your life in the way that you imagined. Of course, this has a lot to do with the **WHY**.

The WHY

The goal results from the **WHY**. If someone has no **WHY** he or she may be able to get going, but as soon as the first obstacle is met, he or she will give up again. And believe me, there will be obstacles...!

If the WHY is big enough, the HOW is not a problem.

We are pioneers in a certain area at a time when many people have not yet realized how important this opportunity is.

In his book "Questions are the Answers," Allan Pease writes:

"The business plan is only the solution to pains your prospects want to avoid or to gains they want to make."

"When you show only solutions without having first uncovered a person's Primary Motivating Factor, the prospect is likely to be swept up in the enthusiasm of it all but not be emotionally motivated."

And:

"There is no point in showing the plan until you have uncovered a prospect's Primary Motivating Factor and have him fired up about it. (...) Find Hot Buttons and press them and building your network will be simple."

In other words it is a question of the order: If you make the problem known **first** and **then** offer a solution, you are in a much better position that if you offer a solution and then start arguing about the problem!

If I offer a top manager an additional income I probably have no chance of getting him interested. He probably already earns so much money that he has another problem: not having the time to spend it. We also know that anyone who earns enough will also have a great deal of responsibility for turnover and risk, and will have to put money before time. We can assume that he only knows his family from the pictures on his desk anyway, and therefore certainly won't want another job on his plate. But what if I talk to him about the value of time and then ask him at a suitable opportunity. *"What would you say if I show you a way of building up a second source of income within a few years that would then become your primary source of income without having to give up the security of your current job?"*

Nor does it make sense to talk to people who have not yet discovered their true dreams and currently only have small goals about a monthly income of 20,000 to 30,000 Euros. You can see how important it is to discover the **WHY**. My motto is:

"Discover what your partner wants and help them to achieve it."

If you really do this, **you can't go wrong.**

Even as I am writing this, it occurs to me how simple the whole thing really is. Or at least appears to be. Nevertheless, I can tell you that this is exactly where the majority of mistakes are made. I know from experience that this point is neglected during the majority of conversations. New partners are particularly guilty of pouring out information about their friends without having the slightest idea what the **other** person wants.

"Discover..." – that is what it's all about. *"Discover..."* means: *"Listen! What do they want – and how does my offer suit their goal?"* In our starter training, we deal with the various **WHYs**, and I would like to like to mention a few more examples at this point:

The value of time

Several years ago, I was with Wissi at a "Career Forum" in Frankfurt. This was a seminar with many well-known speakers. I often visit seminars like this in order to obtain input. Jörg Löhr was one of the speakers, and when we came in after the break, he asked us to reach beneath our chairs. Everyone found a 100 cm tape measure taped to the underside of their chair.

"Now tear off your age," he asked. I was 46 at the time. *"Then tear it off at 79 if you are a woman and 72 if you are a man. We were slowly starting to realize what he was doing..., and then tear off another three segments if you don't do any sport. And two more if you smoke. Then see what's left!"*

This exercise touched me and I often tell the story. I usually add, *"If you are satisfied and happy with what you are doing, then stay that way. But if you are not, change it. And do it now. If you don't do it now, when will you do it? Time is too precious."*

Time is your most precious possession. You know from my story that this was the main issue as far as I was concerned, and I know from my team that the possibility of building up a stable income that will eventually give me more leisure time is also the point at which most interest is shown. I think that it is a considerable luxury to be able to go for a walk and have the time to do the things that are important to me when other people have to work.

Personality development

Referral marketing is the best form of personal development training that has ever existed. The capabilities that you need in referral marketing are the same ones that you generally need in order to be a better person. This means that anything that you learn here, you will need for the rest of your life anyway.

Dirk Jakob expressed it so beautifully: **BE – DO – HAVE**. It is a question of what we **ARE**, and not what we **SAY**. I am certain that 95 % of communication between human beings is unconscious. It doesn't matter **what** you say if it doesn't correspond with your real attitude. This is the reason why conviction is so important.

> **"What you ARE is speaking so loudly that I cannot hear what you are SAYING!"**

This is why mechanically reciting your presentation perfectly is pointless if you don't tell an honest story, for example. Experienced Networkers know this. And believe me, it is better if you say with all honesty, *"I was unexpectedly released by my company after 20 years. It affected me a great deal in the beginning. I can no longer pay the installments for my house and at the moment, I have to watch my spending. But now I can see the light at the end of the tunnel. I have found ways of building up an existence without taking risks…."*

It will give you more credibility and success than driving around in a sports car and flashing your money around and being "ultra-cool" about it by saying things like *"My wife's doing something or other - you can take a look at it yourself if you wish."*

You will be neither authentic nor particularly successful if you take this attitude. As I said, it is not a question of what we **SAY**, but what we **ARE**. The people around us notice this and it is not unusual for friends to become curious when they spot the change.

Our outward appearance also plays a part. "The first impression counts." Several years ago, I read a sentence that said something like this:

"You must make your outward appearance so attractive that people, at first glance, look at you long enough to recognise that you are somebody extraordinary."

For this reason, we have to ask ourselves the following question right from the start: Am I making the impression that I wish to make on other people? One thing is perfectly clear: We cannot **NOT** make an impression! We always make an impression, either positive or negative. This goes without saying.

So many people in my group have said, *"Even if I didn't get a cheque, it would still have been worth while because I have learned so much…."*

In order to encourage personality development, we regularly provide inspirational and motivational seminars. We hire renowned, wellknown trainers whose seminars normally cost between 100 and 1,000 Euros per participant. Because of our group size of 500 to 2,500 participants and the fact that we do not have to work in a profit-oriented way, we can keep costs to a minimum.

In my seminars, I like to talk about the different types of belief or trust that is necessary in our business. The first belief is in your sponsor. If you do not trust the person who is introducing the business to you, or if you do not find the person sympathetic, nothing will come of it anyway.

The next is belief in the products, which is extremely important. So important that I will pick up the topic again later. The most important belief – and this is the one that is the most difficult – is believing in yourself. This is one reason why we offer personal development seminars. The personal development of our partners is extremely important to us. You can visit a top seminar on personality development with us for no more than 15 or 20 Euros. The opportunity to take part in seminars like this at these prices is probably unique.

Apart from the fact that these seminars are good for our conviction and our own gut instinct, we can also invite our family and friends to them and give them a small insight into our team. Believe me, this will always be worthwhile. Our guests believe that it is something special. Here is some enthusiastic feedback from our internal member's area on the Internet that we received after our seminar.

Hello Dear Upline,
My Downline and I would like to thank you for this brilliant seminar. Not everyone can afford 980 Euros, and we are grateful for being given this opportunity for 15 Euros. It was one of the best seminars ever. I am wild with excitement! I am happy because my new member started on the same evening and the first information meetings with my Downline were arranged on Sunday.
Kind regards, Tanja K.

Hello Gabi,
We were in Pforzheim with 17 nice people today. Half of them are not in our Downline. Some of them have already decided to become our partners.
Kind Regards, Ingrid + Michael

Dear Dirk,
Firstly, I would like to thank you for this wonderful day. The day gave me something quite special. My two children, Moritz (16) and Lea (15) were with me, were laughing and clapping and really enjoyed themselves. I couldn't have been given a better present. Thank you for the bottom of my heart!
Regards, Annette P.

Do something meaningful

What if someone did not need money and had all the time in the world? And what if he or she still wanted to do something meaningful?

The book "Your First Year in Network Marketing" by Mark and Rene Reid Yarnell spurred me forward:

> **"Networking is not a job, but a way of doing the things that you want to do in your life."**

The Yarnells have devoted themselves to Africa. They write in their book that there are 5 million helpless Aids orphans in Africa, *"Some things are actually more important than becoming excessively rich. And we in Network Marketing have the time and the money to change things."*

And what about me? I am past 50 now and I have reached all my goals, especially the one of being able to stop working at 50. There were a few months when I did not really know what my goal was? For afew weeks I was thinking that I did not need any more goals. But this only lasted for a few weeks… today, I now have a clear vision!

One thing I feel very strongly about is that I want to commit myself to helping this wonderful industry achieve the reputation it deserves! We have so many problems in our society today which I don´t care to mention. And I agree with a book I recently read which indicates that: *Network Marketing is one of the few opportunities where people of the 21st. Century can determine their income and be independent!*

This is why, when Dirk Jacob suggested to me to found the association Networker for Humanity, short NfH e.V. he was able to count on me immediately. Apart from its humanitarian goals, this association intends to strengthen our entire industry. Through my work for the NfH e.V., I am able to have close contact with the founder members and leaders of other companies which facilitates the exchange of ideas at the highest level, something which is unprecedented. There are regular NfH events. These are wonderful highlights were we learn from the best in the industry and I am sure that together we can accomplish great things. I cannot recommend more these events for those who want to get a better picture of the networking industry. On their website www.nfh-ev.de, you will find an overview of the projects they support. Furthermore, you find a calendar of events and if you think about becoming a member, there is an online application form.

If everyone does the little bit that he or she can, we will all move a long way forward.

Have you heard the story about the starfish?

"A boy went to the beach with his grandfather. When they got there, they saw that thousands and thousands of starfish had been washed up and were now drying out in the sun. The boy started to throw one starfish after the other back into the water. The grandfather said, 'Boy, what are you doing? There are so many that it won't make any difference if you throw a few back.' The boy picked up a starfish, showed it to his grandfather and said, 'It makes a difference to him.'"

Being among people

Have you any idea how many people go out in the evening in the hope of having a sensible conversation somewhere? (I know, I was single for eight years.)

Many people are bored. Even on Majorca many people live in their villas, but are extremely bored because they do not know what to do with their time. Referral Marketing is an opportunity to get together with a wide variety of people, and particularly with special people. Anyone who has been to an event knows what I mean.

Open people who have learned what the prejudice of "Putting opinion before knowledge" means and are open-minded to new things. Many have already leaped over one hurdle and expanded their horizons.

Increasing your pension

I recently read that only 1 % of people over 65 have adequate financial cover. What about the others?

Now, 3 % work, 4 % have an adequate pension, 29 % are already dead and 63 % are dependent on welfare, their friends or their families. Do you think that the situation will be any different in the future?

Anyone who is under fifty and thinks that their pension will give them a dignified existence is wrong.

Referral Marketing is now at the right place at the right time. You will remember Don Failla's question, *"Imagine that the installments for your house and your car were paid and you were still paid 5000 marks per month, regardless of whether you got out of bed or not. How much money would you need in the bank to receive this amount in interest?"*

I have just worked it out: With 5 % interest, which is quite high, you would need 1,200,000 Euros!

With regard to the small "pension gaps," I would like to take up this example again: What do you think you would have to save and have in the bank in order to receive just 400 Euros per month in interest? In other words, an amount that would nicely bump up many a pension? The figure is still 96,000 Euros. How many people can put 96,000 Euros in the bank nowadays? On the other hand, how easy is it to achieve 400 Euros of stable income in Referral Marketing!

More housekeeping money

I am extremely proud of the many mothers (some of them single mothers) who have been able to boost their budgets thanks to their work in Referral Marketing. At one stage, I too had to think about whether the monthly budget would run to a new pair of trainers (for those of you who don't have children, you would not believe how expensive they are), or whether I would have to wait until the following month. My son sometimes outgrew everything too quickly as far as I was concerned, too.

200 or 300 extra Euros per month can make a big difference. And once you have achieved this goal, you can increase it, of course. I am extremely glad that I no longer have to do any cleaning and ironing. At this point, I would like to show you a letter that I received recently. It came from a single mother with six children:

> Dear Gabi,
>
> I simply have to write a few lines to tell you what is on my mind.
>
> I know that this business is the best way for me to give my children and me a better future. My children finally have a perspective in their lives. Also because they can see how I have developed as a person.
>
> I did not think that we had a future, and even if we did, I thought that we would be dependent on welfare, and that my children would become the next generation of welfare recipients.
>
> Before my ex left me three years ago (25 years of marriage and six children who are now 28, 17, 15, 13, 10 and 7), we already didn't have enough money to cope with any additional requirements, let alone go on holiday with the children.
>
> After the separation, it became really tight. Buying a second pair of trainers for doing sport at school was simply impossible. And it hurt me a great deal to see what my children had to go without.

Social Services informed me that I would have to get rid of my car if I wanted to receive continuous payments. Shopping would then become a task that could take all day. They also demanded a collateral mortgage on our little old house. And if I refused to do this, I would have to do community work.
So I went without the payments from Social Services and cleaned rooms and baths for 5 Euros an hour. When I got home at about three in the afternoon, I carried on. I was a physical and mental wreck and had to neglect my children.
Whilst I was this low ebb, I heard about your company and knew that I could build myself a future and remain independent.
I know that I am not the only one to be in this situation. I just want to show other people in similar situations that they can also do what a single mother with five children has done. I am bringing the three youngest (aged 7, 10 and 13) with me to the summer party in order to show them what life still has in store for us, and as a reward for that fact that I can build up my business in peace and quiet in the evenings.
Thank you for believing in us – also on behalf of my Downline.
We appreciate your honest interest in our growth and do not take it for granted.
Kind regards from the Saarlanders…
G.

This letter contains many reasons for the business idea. I put letters like this in my "success file," which is already quite thick. And every time I feel down, I take a look in it… You cannot believe how much it motivates me. Letters like this are the reason why I do not want to, or cannot, give it up. I would like you to receive letters like this too.

What would be your reason? What is your personal WHY?

- Do you want to combine a career and a family?
- Do you simply want to go shopping without being answerable to anyone?
- Are you not particularly fond of your boss and your work?
- Do you already earn enough money but do not have the time to spend it or the time for family, children, church, sport or whatever…?
- Is your current job leading you to a heart attack or a divorce?
- Would you like more appreciation for the work that you are doing?
- Is your company close to collapse?
- Are you a woman who has trained in a good profession but have no chance of making a career of it because of children and family.
- Or would you just like to make money without limits?

How Network Marketing Developed

When I started out in this industry eleven years ago, Network Marketing was a kind of stepchild, and I am extremely glad that this business area is enjoying an increasing amount of acceptance, which is mainly attributable to the general economic situation. This is particularly true of the type of pure Referral Marketing that we practise.

Network Marketing has been enjoying continuous growth for many years. Whereas there was just a handful of companies 30 years ago, now there are an estimated 100 companies in Germany alone. Companies try time and time again to establish themselves on the market, but underestimate how difficult it is to set up a functioning Network Marketing company and then disappear again. The market is still dominated by the Americans, simply because they have the most experience. Franchising was laughed at in the early days, but nowadays our economy can hardly exist without it. In the pedestrian areas and shopping malls of the big cities, you always see the same shops – most of these are franchise companies. Franchising forced its way through and is still a model of success that is associated with extremely high start-up and running costs, which prevents it from being available to everyone.

Network Marketing is different. However, the principle is similar: The model business has to be started up during a trial period, the procedure is standardised and should ideally be documented, leaving nothing in the way of further duplication. But the horrendous costs and fees are nowhere to be seen.

Network Marketing is worthwhile for both financial and social reasons. Let's think about why this is the case.

A job for life

A job for life, as we know it, is a "discontinued product." Researchers (and "Mister Strachowitz" with whom I have just carried out a telephone interview) think that future companies will just consist of a core team, and freelancers and legally independent small companies will carry out all other work. In future, employees will no longer be individuals but small companies.

First indications: Promotion of "Me inc." or bridging money for people who want to make themselves self-employed. However, the resulting income is linear, which means that the self-employed person works hour after hour and is paid a certain fee. This person's income is restricted by the time factor and cannot be duplicated, which is where Network Marketing has the decisive advantage.

In properly organized Network Marketing, and particularly Referral Marketing, you duplicate your sales partners yourself and achieve income and volumes that could never be achieved using the linear method.

Insolvencies

Forty thousand insolvencies were registered in Germany in 2003. These included both hopeful new starters and companies that were rich in tradition. This is all the result of the restrictive credit policies of the banks and exuberant corporate cost structures, caused by tax and contribution policy, and extremely high labour costs. The German model that our parents knew doesn't work any more.

Pensions at risk

The situation is made more difficult because of the demographic situation – people are living for longer and are breaking the bank as far as pensions are concerned. I can be brief with this topic – you can read the details in the newspapers.

Karl Pilsl, a well-known economic researcher, wrote in his book "The Natural Strategy" ("Die Naturkonforme Strategie"):

"If a young man or young woman went into Network Marketing after high school and built up their network as 'on-the-job training' on the side full time with a good and reliable company whilst they were studying (e.g. for four years), they would graduate after four years with the right to a pension."

Wouldn't that be great? I have had an increasing number of extremely young people at the training recently and envy them for having the opportunity to learn about the business at this young age.

Secondary income

Everyone decides for themselve whether they want to make an additional income out of it or attempt to achieve a substantial passive income. I just want you to know that these are two completely different things. The difference lies in the number of contacts that we make and on the other hand on the speed. Do you remember the example of the petrol station? If everybody just sponsors one other person per month? What do you think is the result instead of the 4,096 people, when we take more time and inform a friend only every second month? And this keeps on duplicating? What would you guess spontaneously? Most people answer: **well, maybe half!** And that´s what blew me almost away, after seven years of doing the business. It is not just half, it is just 64 instaed of 4,096. Although I am doing the same, just a little bit **later**! By the way this is another example, which shows that this industry has completely different rules and that you actually can forget what you have learned before. This is why it is so important to stick to a certain system and do not try to re-invent the wheel.

Occasionally I hear the argument *"Not everyone earns big money!"* This is true. Apart from the one small difference, in this case everyone **has** the opportunity to be successful and earn big money, but not everyone is prepared to do the things that are necessary to get it.

At this point, I would also like to emphasise that I find it just as honourable for a person to want to plug a 250 Euro pension gap as to want to achieve financial independence. In 2005 I read an article of a big German newspaper that 43% of all households do not even have 100 Euro available income at their disposal. This means that the many typical small incomes of several hundred Euro as an additional income can often represent a doubling or multiplication of the available income. Jim Rohn – a well-known philosopher and Network Trainer – writes about the "magic of secondary income."

> **"If someone earns an additional 1,000 Euro in their profession, no-one is interested. But the possibility of earning 1,000 Euro on the side is a major attraction."**

Have you ever thought about how brilliant the possibility is of building up a secondary income without having to give up the security of your existing income?

Definition of terms

Emotions play an important part in our business. In spite of this, for the sake of better understanding I would simply like to explain a few terms that (if you will forgive me) are somewhat dull.

Whilst I am writing this paragraph, I feel the need for explanation – Network Marketing, Referral Marketing, MLM, Multi-Level, Direct Selling, and Direct Marketing – how can I explain these terms in a way that is correct and understood by everyone?

I could use the help of "Mister Strachowitz" at this point – after all, he has 30 years of experience in our industry and is therefore the oldest "network dinosaur" that I know.

We have talked about this topic at length and the only thing that I know is that no definition is 100 % correct. Everyone understands too many different terms in different ways. And some of them are so worn-out (perhaps by representatives of the industry who are not particularly serious) that it is better to stop using them (e.g. multilevel selling). For this reason, new terms such as Access Marketing or Team Marketing will be invented time and time again. Perhaps a new expression will have been invented by the time you read this book. One thing that is certain, Network Marketing means "accumulation of small sales."

I have now spent half of the afternoon thinking about how I should finish off the topic in a fruitful way. In the end, I decided to give you **our** definition of the term. By "us," I mean *"GabisteinerTEAM."* This is an amalgamation of a large number of managers and partners from my team who live according to a common philosophy. I will tell you about this in detail elsewhere. I must point out that the following explanations do not claim to be **generally** applicable.

Network Marketing

We understand Network Marketing to be a special form of selling, for moving goods from the manufacturer to the customer by bypassing the local retailer. In Network Marketing, commission is paid for assistance in setting up the organization (this means sponsoring other people who are doing the same thing), usually at different levels. For this reason, it is also referred to as "Multi-Level Marketing."

The majority of companies that have chosen "Multi-Level Marketing" as their selling method are a hybrid of Network Marketing and direct selling, i.e. part of the commission goes into direct selling.

Direct Selling

In direct selling an adviser purchases discounted goods and then sells them on to the end consumer. The adviser earns the difference between the purchase price and the selling price.

Qualification Volumes

The quantity of a company's products that an adviser has to sell each month in order to receive the commission for support in setting up the organization (i.e. the networking) is called the qualification volume.

Sponsoring

The acquisition of new partners is called "sponsoring" (supporting). The person who has successfully introduced you to the opportunity is your sponsor.

Upline

Your Uplines are the people who were in the line before you, i.e. your sponsor, your sponsor's sponsor etc., right up to the top.

Downline

Your Downline consists of all the people that you have introduced to the business and the people who they have introduced to the business and so on. The amount of your monthly income is determined by the size and activity of your Downline.

Sideline

Your Sideline consists of all those people who work like you in the same partner company but are in neither your Upline nor your Downline

Referral Marketing

Pure Referral Marketing is understood to be a type of Network Marketing in which the qualification is usually so low that anyone can achieve it by merely covering his or her personal requirements (think back to the refuelling!). Anyone can build up his or her own network by recommending this concept or idea without having to sell products to end consumers and bring in money. And although no one has to sell products, the company achieves turnover that allows it to pay out commission at various levels.

The great thing about this pure form of Referral Marketing is that you can build up a stable and above all "stable" or automatic income with a product that you need anyway and consume every month. You will find a few example figures in the following chapter.

You are now aware of the importance of explaining the differences between retailing, direct selling and referral marketing to our potential new starters. In our system, we make a distinction between pre-information and catalogue information. Both systems are at the core of our presentation. And as I have said, it is about the framework – the content depends on the target audience. For example, if I were talking to someone who knows direct selling and is absolutely determined not to do it any longer, I would go into this issue in somewhat more detail than I would with a person who was coming into contact with our industry for the first time.

Pre-information

At this point I would explicitly like to mention that I only provide this pre-information if I have made a "good connection" with the interested party. Personally, I prefer talking to interested parties on a one-to-one basis so that

I can focus all of my attention on this person. I notice where additional explanation is required and I can stop immediately if I notice a negative reaction from my conversation partner.

I often provide assistance when my partner is talking to his/her friends. Here too it is extremely important to "level the ground" first. I tell my story, ask questions and show an interest in the people.

During the pre-information phase, it is just a case of giving an **example** as to how referral marketing generally works. This means keeping the figures rounded. The aim is to explain referral marketing in a simple way and discovering whether our conversation partner is generally interested in this method of selling. In the pre-information phase we can explain how products are moved with the aid of **simple** drawings (remember the framework from the filling station story).

How are products moved?

If you need a certain product, you usually go to the shop. This means using a retailer.

Retailer

Retail sales often use a traditional distribution system.

Depending on the product the manufacturer retains 15 to 30%. The product is sent to a wholesaler, who buys huge quantities of the product and then sells to a retailer. The retailer puts it on the shelf for the end customer. You as a customer pay the normal shop price, i.e. 100%.

As you can see in the chart, the distribution chain eats up the majority of the money (i.e. 70%) in advertising costs, the shop rent and the employee´s salaries, regardless of whether turnover increases or falls. Many self-employed people understand the problem and are suffering because of it – particularly at present.

Direct selling

Another way of moving goods to consumers is direct selling. In direct selling an independent consultant purchases discounted goods from the manufacturer. He usually pays for the goods himself and keeps them in stock. His revenue comes from the fact that he looks for his own customers, who purchase the goods. The independent consultant usually receives between 30% and 50% discount.

However, it is important for the independent consultant to be a proper product expert so that he can give comprehensive advice. He must know all the details about the products. He has to start looking for customers again every month so that he can sell his products. He has to keep finding new customers. His income is restricted by the time factor. In plain text, this means he exchanges time for money.

Referral Marketing

Referral Marketing works completely different. The products go straight from the manufacturer to the end consumer.

And whether you are aware of it or not: you are already doing it. However you probably aren´t being paid for it. When you see a great movie in the cinema or read an exciting book, when you have eaten well in a restaurant or have been satisfied with a product, it is the most normal thing in the world to tell your friends and acquaintances about it. You do this without background knowledge and **without training!**

I always like to ask new partners the following question: "Have you ever recommended a two-hour cinema film and told people about it?" All of them have, of course.... Then I ask "***How long did you train for it?***"

If you recommend a cinema film to someone, you probably don't know all of the actors, nor know anything about the production costs, nor who was responsible for the make-up or the script. You enjoyed the film, you laughed or you were touched. And then you tell your friends. And because of your recommendation, either they go or they don't. Are you thinking about this?

See what I mean? I would find it just as unnatural to invite my friends to the local pub to tell them about the content of the film – or even print a flyer or brochure containing the words *"I have seen a great film."*

As far as I am concerned, a "word-of-mouth" referral should be exactly that, a recommendation to a friend, therefore totally credible and the most natural thing in the world. Among other things, we recommend several independent tools for the purpose of reinforcement and awakening interest These tools are suitable for anyone and have various advantages:

1. They provide an **unbiased** explanation of the industry and are company-independent.
2. Anyone can pass on a tool, which means that this tool can be fully duplicated – one of the most important issues in Referral Marketing.

3. You can put even the shyest of people in a position to inform someone about the nature of Referral Marketing after a brief conversation or after telling their story.

You can obtain a list of current tools from your sponsor during your starter *training, or download it from your member area. The partners of the GabisteinerTEAM* will find all the important reinforcement information that they need on this web site.

Support from the Upline

Tom Schreiter, who is a well-known Networking author, prefers the "two-to-one conversation." This means that your sponsor accompanies you to begin with. He writes:

"The solution is the two-to-one presentation. Since our self-consciousness does not suffer from continuous rejection, we are in a position to give good presentations. Your partners just need to set the date, lean back and enjoy the show. In this way the system helps to reduce fear, since the partner does not have to speak to his interested parties alone."

And continues:

"We ask him to make a few appointments and just observe how we acquire new partners for our group. We build our group up whilst he is watching."

I have spent a long time wondering whether this statement contains a contradiction. Should the new partner carry out his own conversations or should the sponsor do it for him? Here too I can say from experience that it depends on the person we are dealing with. The result is important. And I know from experience that the results are better with new partners if they use their sponsors. The sponsor already has more experience and is (usually) in a position to show you a statement. Personally, I always show new partners the statements from my first few months when I take part in a conversation. I would therefore strongly advise you to obtain the statements from your Upline if you don't have any already. Our interested parties do not usually understand how the duplication works.

Regarding this subject I would like to quote Robert Pauly, a renowned coach in Network Marketing:
...basically there are two different ways of going about it: the first way is relatively simple and rarely works, yet is used by many networt companies. It consists of teaching as many people as possible together in a theory class (often with a built-in "rah-rah" motivational component) to encourage lots of people to get started at the same time and produce a high turnover, something which hardly ever happens. The second way works quite well but it is significantly more time consuming. For this reason, few people use it. This working method consists of focusing on one person alone and building up a team together with that person...

I would like to tell you what I do: If a new partner is interested, I want them to see that it works as quickly as possible. I know about the 72-hour rule, and it is extremely reliable. The rule is generally applicable to all areas of life and says that anything that we do not start within 72 hours rapidly reduces the chance of it ever being done.

Everyone has at least one friend who may want to get involved. In an ideal case, the friend should be there at the starter training. You could invite them as follows:

"Hello Bernard, it's Anne. I'm sitting here with Gabi and I have heard about a brilliant way of earning the money to pay the instalments on my car. On Friday, I am going to a starter training session and I would like you to come along too. Your opinion is important to me and if you're interested we could start doing it together."

If Bernard doesn't have any time on Friday, the date can be moved to one that Bernard suggests. Do you think that your best friend would deny you this request? And if Bernard sees an opportunity for himself and we offer him actual assistance (which he has just experienced himself by helping Anne) and takes part, the next conversation can take place with Bernard's sister Christine. Perhaps Anne will watch again, or perhaps she would rather carry out the conversation herself. It doesn't really matter.

After all, the aim is to put new partners in a position to explain the opportunity on their own after a certain time.

"You don't get muscles by watching the trainer doing bodybuilding."

I therefore recommend:

> **"Don't do anything for someone that they can do for themselves."**

The main thing is we are there from the start as the sponsor and are able to convince ourselves that they can "do it." It is certainly true that someone who received active support right from the start is much more successful than someone who was sponsored and left to his or her own devices.

There is also another thing that I see confirmed repeatedly. There is no better advantage or motivation that having someone behind you right from the start.

When I started in April, I sponsored seven people, most of whom were family members. These included Wissi, our grandfather, my mother and a few friends. In May, we added Lissy Schütt-Nothdurft to the list. At the time, we all took a walk together every morning for an hour. Lissy was immediately enthusiastic. She spontaneously wanted to pass on her enthusiasm to her husband Werner, albeit not very diplomatically. Now, Werner is a qualified engineer and father who is extremely set in his ways – and Lissy was going into raptures at top volume *"Fantastic products... and you don't have to sell anything... it's brilliant...,"* whereupon Werner just came out with a single comment *"And you believe all that?"*

They argued about it for a long time, but Lissy remained just as enthusiastic, and for a month or so, we carried out "Undercover Networking." We spoke on the phone at least five times a day, and every time Werner came home from work, Lissy whispered quietly into the telephone *"I' ll have to hang up, Werner is coming."*

Four weeks later I spoke to him and he agreed. Eighteen months further down the line he was overjoyed to be able to quit his job and spend his time at home with his family as "Free Werner."

But back to Lissy – she motivated me no end. In May I spent two weeks in the Dominican Republic (that was the only place where you still didn't have to pay for kids up to the age of 14 – my budget was still small in those days) during the Whit holidays.

During this time, Lissy sponsored Isolde and Ottmar from Ettlingen, and when I returned from the holiday a call from the "Fast Talk" information system of our partner company confirmed that I had achieved "Bronze" status. This is the first management level in our compensation plan. I almost fell over when I realized what Lissy and her group and my other "three Annes" had achieved in my absence. Now I was certain that the business worked.

Shortly afterwards I also knew that the products worked – there was a rumor that I had not been to the Dominican Republic at all, but had secretly been for a face lift.

Isolde and Ottmar were to Lissy what Lissy had been to me: Both were independently active at this time. Isolde had a translating agency that brought in an above-average income. The thing that she did not have was time. Ottmar had been self-employed in the building trade for ten years. In spite of working long hours, he earned less than a salaried employee, and was investing more and more money in building up his business. The financial situation was visibly becoming more difficult and his company was starting to lose serious amounts of money. When he accepted the inevitable consequences in 1999 and shut down the company, he found himself in an extremely tricky financial situation. This was precisely the moment in time when the two of them discovered Referral Marketing; they immediately recognized the marvellous opportunity and got going extremely quickly. Nowadays they "work" together and are enjoying their newly found freedom and independence.

My basic recommendation to you is the following:

Look for a person among your closest friends whose company you enjoy and who is willing to start up with you. It is a priceless advantage to have a motivator at your side right from the start. The best motivator is speed.

Now that I think back on it, the way in which we worked was exactly as described in the literature as being a successful way of operating:

Within the team, we worked the **depth**. Here is a brief explanation of what I mean by "depth" and "width":

Everyone who I sponsor personally is referred to as **width**. For example, if I directly sponsor five people (five Annes), this is my first level, my Firstline. If my Firstline Anne sponsors Bernard, and he sponsors Christine, who then sponsors David, my organization is four levels **deep**.

Now, Lissy was nowhere near ready to speak in front of people (not that you'd notice it nowadays…), and she was more housebound that I because of her two children. However, she was extremely good at "contacting." We did some training together with her people and I explained the system, and we started in exactly the same way with Isolde and Ottmar in Ettlingen and also with their team partners. We had a lot of fun together!

Observant readers will have noticed that I briefly went off the subject to spice it up with some interesting stories. This is exactly what I do in practice. Remember: **Networkers are storytellers.**

Now let's examine the contents of the pre-information. In our example, you have spoken to Anne, because you have a certain amount of previous knowledge, and I can assume that you are interested in the subject. If someone just asks me what **I** am doing, I use the "I" form in the start information, not the "you" form, i.e. I have spoken to Anne. This means that **I** am talking about myself, and I do not have the feeling that I am pushing the other person, nor does the other person feel that they are being pushed.

Let's take the example of a product. Just like you, your friend Anne buys the product straight from the manufacturer and gives her customer number as a reference so that the manufacturer knows where the order has originated. If Anne is also enthusiastic about this product, she is certain to talk to other people about it. Assume in the example that Anne tells her friend Bernard about it (among others) and as a result, he orders directly from the company.

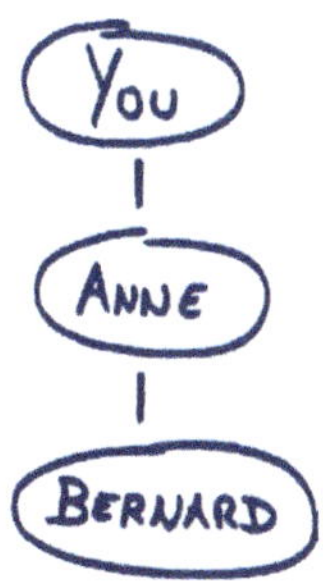

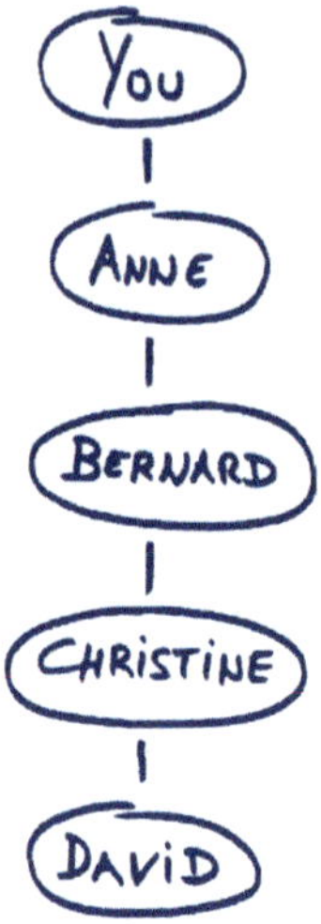

Bernard then tells his colleague Christine about it, and Christine tells David with the same result. David now also orders his products from the manufacturer and gives Christine's customer number as a reference.

We can use David's order to show exactly how commission is paid in Referral Marketing. The 40 % that would all go to the adviser for his/her advisory work in direct selling in our example is divided among several people in Referral Marketing.

In our example, it would look as follows: Christine receives 10 %, Bernard receives 10 %, Anne receives 10 % and you receive 10 % from David's order. After all, you started this referral chain in the first place!

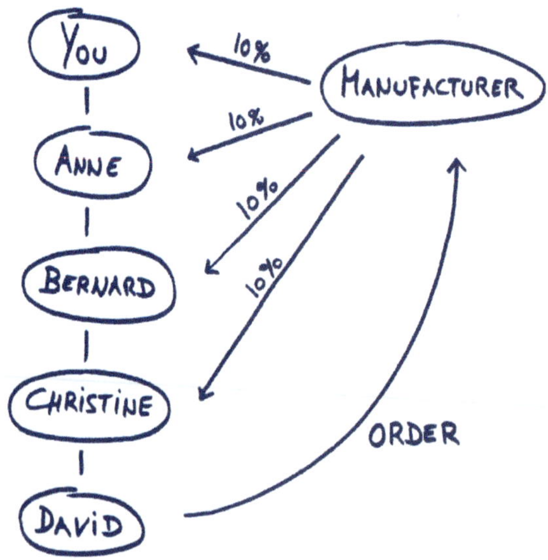

At this point, it is extremely important to me that my discussion partner understands what I am saying. I would like my partner to nod, at least internally; otherwise, I do not carry on to the bigger numbers. There is an important reason for this: A discussion partner who finds the system good with one Anne will not have a problem with several Annes!

During my discussion with my brother, I knew that he was having financial problems with his company and that he and his partner would soon be unable to pay out a salary. With two small children and installments on the house, this becomes a problem, particularly when you are self-employed and cannot fall back on the security of our welfare system.

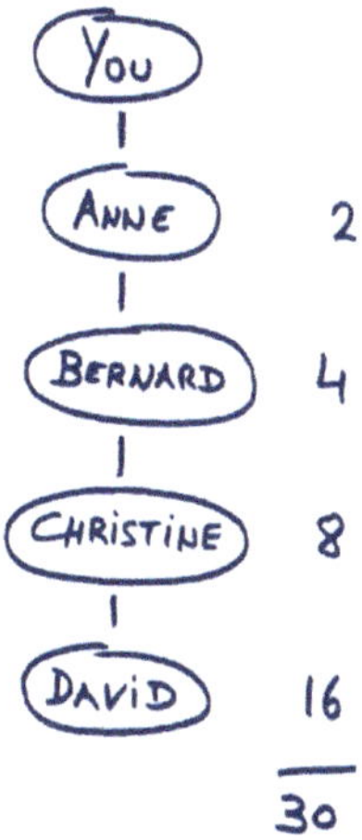

I told him that we could work with two, three five or more Annes, depending on my discussion partner's target. "Anne" stands for people who we talk to directly. For this reason, we have used the first letter of the alphabet. I said to my brother: *"YOUR target is somewhat bigger. One Anne is not enough. I will therefore show you the power of multiplication. Let us just calculate with two, three or five Annes. If you have two Annes, both of whom sponsor two Bernards, who then sponsor two Christines, and they sponsor two Davids, we now have a total of 30 consumers.*

I will now demonstrate the power of multiplication by taking just one additional Anne. If we duplicated (repeated) this through to the Davids it would look as follows."

"That makes exactly 90 additional consumers ("refuellers"), although we have only sponsored one additional Anne. You have to set up a new business, so your target is somewhat bigger, which is why we will both work with five people."

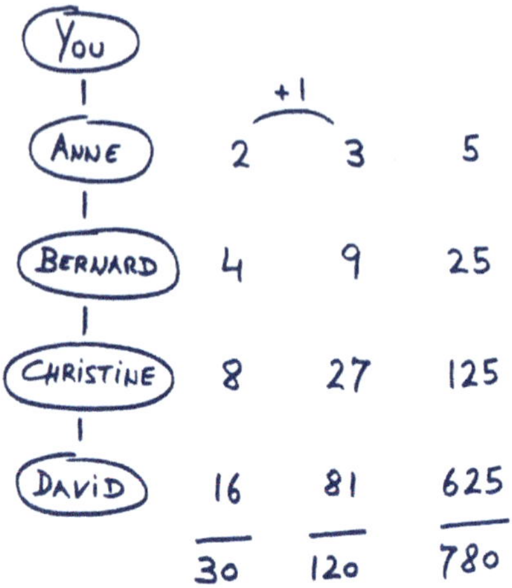

"780 people who purchase and consume 100 Euros' worth of goods (100 Euros makes the calculating easier). No-one needs to sell products or collect money. This means that your group would achieve total turnover of 78,000 Euros."

My brother almost fell off his chair when I asked him *"How much would you receive in the example?"*

780 x 100 € = 78,000 €

OF WHICH 10 % = 7800 € PER MONTH

With the 10 % in the example, that would be 7800 Euros! The only other question that I ask is:

"Is this interesting enough for you to want to know more?"

Andy wanted to. Of course, afterwards I also asked him the questions that you are already familiar with from the filling station example: *"In your opinion, what makes a product suitable for this selling method?"* Remember? It has to be a consumable product, usable for anybody, and be in a fashionable market.

Now came an extremely important moment: We had simplified our presentation even more, and the catalogue that we had designed together with the company management had just arrived on the market. I used Andy as a sort of "guinea pig;" in those days he was a shy person (which has changed in the meantime). And I knew that he was aware that I could do it, but it was also clear to **me** that I would have to keep the catalogue presentation simple enough to make **him** believe that he could do it himself.

The catalogue information is no more than word-of-mouth referral. From person to person. Pure and simple. I took the catalogue and went through the quick version with him.

Then we talked about the compensation plan and suggested an initial order to him. Since my brother was not a fan of food supplements (something else that has also changed in the meantime), I did not say anything more about the products. When I started out back in 1999, one of the most important things to me was the existence of Anne Simons' OPC book. The book is independent and available in any bookstore. One thing was immediately clear to me in those days: If somebody outside the company has written a book and describes OPC as the "anti-aging vitamin of the century," this is the ultimate proof and no longer requires discussion! For this reason, I try to have the book in my hand during every conversation so that I can show it to my conversation partner. I think that new starters need some kind of "proof" and I believe that the quota of new starters is considerably higher if the book is used. This is why I cannot understand why this "quota booster" is not used more frequently....

Finally, full of hope, I asked my brother the most important question: *"Do you think that you can also do it?"* To which he replied *"Of course, no problem!"* I could have jumped for joy! I had achieved two extremely important things. All new starters have two burning questions. The first one is "What's in it for me?" This is a matter of the **WHY**. I knew why he wanted to get involved and incorporated it into the plan. The second question is "Can I do it too?" And he had replied to this with "Yes!" The time that is required to do it also plays an important part as far as the second question is concerned. The new partner must **also** be able to figure out how to fit it into his busy schedule!

Temporal and physical independence is an important reason why I prefer the "One-to-One" conversation. And something else that is important: if we use tools like CD's and books and the interested party has already heard or read some information, not much explanation is required. My interested party is them more likely to imaging having the time to do it and being able to do it.

Michael Strachowitz wrote the following in his article "Das persönliche Rekrutierungsgespräch" ("The personal recruitment conversation"):

"Where is the right place to introduce my own company? You rightly ask yourself. The logical reply is: the place where your company has its headquarters – at your house! Network Marketing is a business that can be run from home. It is precisely this argument that makes it so attractive to many people! And your interested parties should get to know it just like this – as a home-based business!"

I think that it is easy to carry out a personal conversation and pass on catalogue information anywhere and at any time. The catalogues are now available in many languages (partners in my team can also find the forms in the member's area on the Web site under the "Forms" link).

Referral is the best form of advertising

The most effective advertising is word-of-mouth referral by a friend. Nowadays we have "information overkill," which means that we are besieged with as much information every day as a person received in an entire lifetime 100 years ago! In order to protect ourselves from this, I think that we just shut our ears and only open them if… what? What is it that makes our ears prick up? Just think about it.

There are two extremely important points. The first is **curiosity**. Curiosity is the biggest crowd-puller known to man. If we manage to make people curious, they will definitely listen to us!

The second point is the **relationship** that we have with the person who is telling us something. That is the most important factor. I know now that even the best of presentations will not work if the chemistry is wrong. Have you ever bought something from somebody that you don't need, just because the person was likeable? On the other hand, you have probably decided not to buy something that you needed because your gut feeling wasn't right. Do you really believe that purchasing decisions are made for rational reasons?

Our grandfather (87 years old and still looking after himself) recently came out with a classic example: *"I buy my ready meals from 'Bu-Frist'. They are more expensive than 'Schneemann', but they have a nicer driver!"* This is why I see the future in "relationship business," which is far removed from a "business relationship."

The power of duplication

Effectiveness lies in simplicity. The more frequently we manage to create business processes that are easy to duplicate, the more success we will have.

"It is not enough for something to work – it must also be capable of duplication!"

This is an extremely important principle. Being able to duplicate is allimportant. For this reason, you can do yourself and your sponsor a big favour: Don't start thinking about how the brochures could be improved right at the start, or whether something could be modified a little here and there....

In "The 5th Principle" by M.S. Clouse, I found an interesting passage about energy-charged entrepreneurial types making changes to a system:

"The challenge – and it is a challenge – is even if their (or your) method is to improve, it will not work in the long term. Why? Because this business attracts too many energycharged and enthusiastic entrepreneurial personalities. If they all succeed in forcing through their attempts to improve the system, before you know it you won't have a system at all! Ask yourself the question: 'What do I really want?' And if your answer is 'To build something up and live off it for the rest of my life' you should appreciate the simplicity of the system that can be duplicated, and concentrate your entrepreneurial spirit on building up your sales group!"

I like it! And it's extremely important! Believe me, there are several people in your Upline, including me, who are extremely keen for you to be successful. If it is possible to make any long-term improvement, we will give you some training.

This why we have the frame in our system. The frame, or scaffold, means that the contents of the training are identical. Not the words, of course. That is an important prerequisite for being able to give the same training anywhere in Europe (or perhaps even worldwide). But believe me, as soon as your team has reached the size that you want, you will be **extremely** grateful for our training calendar. This is where you will find the dates of training meetings or events throughout Europe in which your partners can participate, regardless of where they live.

Now a few sentences on the practical effect of duplication.

In our example, we carefully calculated up to the "Davids" who were sponsored by the "Christines." In other words, up to the fourth level. However, this doesn't have to be the end. In pure referral marketing, where no money is needed for intensive advisory work, commission can also be paid at deeper levels! You can work out for yourself what happens if the "Davids" show the "Emilys" in their circle of acquaintances this brilliant opportunity, and the "Emilys" show the "Fredericks," the "Fredericks" show the… etc.

I this point I would like to explain one of the most important issues that is occasionally underestimated, in my view, and possibly even neglected as a result. First a quotation from the book written by my sponsor, Don Failla:

"Referral marketing works best if many people simply cover their own needs and supply their own personal environment (friends, family and acquaintances) with products. This considerably more effective that a top salesperson doing it all himself. Millionaire Paul Getty said, 'I would rather have 1 % from work that has been done by 100 people than 100 % of my own.'"

Let's have another look at the figures for double, triple and quintuple duplication:

There were 30, 120 and 780 active partners in our example. With 780 partners in the team, any company would have considerable turnover. But what if each of the 780 active people had a mother who just used the products?

The turnover is no longer 78,000 Euros as was shown in the example, but is now 156,000 Euros! And now imagine that everyone recommended the products to their parents and their grandmothers! In other words, every active person had three customers! That would quadruple the turnover (me + parents + grandmother = four people) = turnover of 312,000 Euros! Can you see how important this component is? The more ambitious of you may even imagine having five or ten customers – I'll leave that to your imagination!

Our company has a product line consisting of more than 60 products, from natural food supplements to wellness, sports and best-aging products. Furthermore, our company provides products, which we use anyway like toothpaste, skin cream and shampoo etc. Not to forget the products that everyone has in his medicine chest. There is something in it for everyone. If every one of these 780 people also bought their skin cream and their toothpaste from the same company as they bought their basic vitamin supply from, would the resulting figure be too high for you?

Let me tell you something:

Our toothpaste (which is extremely good by the way) has 3.1 volume points. Volume points are an artificial currency and are expressed in International Points (IP's), whereby an "IP" is the approximate equivalent of one US dollar. If every active person in my team ordered from the same company, that would currently mean turnover of 90,000 dollars in my entire team. Five percent of this would be a nice monthly salary increase. Fantastic, isn't it? That is Referral Marketing!

And I'll tell you something else:

In my case it doesn't matter any more, but when you start out in Network Marketing, regardless of the company you are with, I can give you a piece of good advice: If certain products you need are available from your company, always buy them from the company, even if they cost a few Euros more. I have already mentioned that quality is of primary importance, particularly when this selling method is being used. And quality comes at a price. **Always**! Particularly with food supplements it is extremely important whether the vitamins have been manufactured artificially or whether they come from fruit or vegetables.

At this point, I must mention that all products that are sold in direct selling generally have to be better than the products in the shops. Since we are moving in a "warm market," i.e. our circle of friends, it would be a catastrophe if the products didn't work. For this reason, the products that are sold in Referral Marketing are all top quality products. Experience has shown that companies that do not sell tip-top quality do not remain on the market for long. Paula Pritchard writes in "Owning yourself":

"When you first start out in the business, order two of every product that your company is marketing: one to try out yourself, and the other to sell".

Now the good news: with us you don't have to resell – this is why one extra product is sufficient!

We do not need a wealth of KNOWLEDGE, but a wealth of EXPERIENCE!

Back then I said to my mother: *"You know my situation – I simply can't achieve my goal with my business, and I have therefore started up a second income with an old-established company that has decades of experience in antiaging products and food supplements. Here is a book to read about 'OPC', which is one of their main products."*

There was no doubt in her mind. On my web site **www.gabisteiner.de** you will find a picture of her at the age of 77 under "The Way" (towards the end). She is living proof that the products AND the business idea work....

However, what if I only had a PRODUCT to offer? If that were the case it would come as no surprise if the person I was talking to considered me to be a salesperson. Don Failla said: *"Most people find out about Network Marketing from friends or acquaintances who offer them certain products to buy. It is therefore no wonder that everyone confuses Network Marketing with direct selling!"*

Particularly since there have been many heated discussions about it and many misunderstandings have occurred, I have devoted an entire chapter to it on my web site:

The chicken, or the egg?

A topic that appears to be conflict-ridden in referral marketing is the question of *"Product or business"* – or *"the chicken, or the egg?"*

It took years for me to finally clarify this matter. I believe that I now understand something about referral marketing. And I can now give you a clear answer: it has nothing to do with the question of "Product or Business" or even having to choose one or the other!

It is not about the chicken or the egg – it's about the whole farm! It's about people. About their goals and their dreams! In our example it's about the door through which people enter the farm.

I have been a complete vitamin addict for more than 30 years. Even as far back as 1972 I was trying to recommend a healthier diet to people with health problems in my circle of family and friends. Do you know what I discovered? *"A typical Swabian would rather die younger than go without his roast beef, potato salad and pasta."* What am I trying to say with this?

My original goal was to give more people a better diet and be able to lead a healthy life. How was I going to achieve this? By convincing them to take vitamins?

I now know that this method is the most difficult one of all. It is also fairly frustrating. Anyone who has tried it will know what I am talking about.

Dr. Joe Rubino writes:
"Network marketing was founded upon the concept of sharing great products with others. In the 1950's and '60's, the marketing spotlight was focused solely on the products themselves. Sales techniques were aimed at convincing customers of the value of particular product after explaining all of its features and benefits. Show an tell was what it was. (…) Show the product. Tell the story. This basic show-and-tell technique worked great in the product-oriented past when there were only a few thousand products in the whole world a consumer could buy. (…) Can you see why, today, the focus cannot be on the products? Sure, there have to be treat products – and there usually are.

(…) But there's more – much more. (And the more is – people. Network marketing is about people. At its best, network marketing is about people getting control of their lives, following their dreams, living their values, AND supporting others to do the same. (…) This is what I am still enthusiastic about – the opportunity to lead a different life. And also having the opportunity to help other people to achieve the same thing."

Even those people who cannot or do not want to sell! According to DonFailla, this applies to 95 % of all people. This is why Referral Marketing is so brilliant – because even the shyest of partners is capable of telling their friends about a better life!

Dr. Rubino continues:
"That's why traditional sales people often fail miserably in MLM. They're so indoctrinated in product, product, product, they lose sight of people, people, PEOPLE! Their focus is in the wrong place – either on the products, or on making themselves a lot of money selling the products. Neither of those reasons are very important to anyone else in the world. What IS important to everyone in the world is relationships. And network marketing is about relationships. Period. That's why the old recruiting theory of 'throw enough mud (i.e., people) against the wall an see what sticks' simply does not work. MLM is not about other people doing it alone and making you a lot of money. Rather, it's about YOU partnering with others to help them realize their dreams. And, it's about those people partnering with more 'others' to help them realize their dreams, and so on."

I see our business in exactly the same way. I am proud of working in the way that we do. Of course, not all of us will be successful, but those people who are not successful will not have broken the bank or lost anything (apart from a few wrinkles), and I am extremely proud of this fact.

I take particular pride in our team. They are great people. They are filled with enthusiasm. They take responsibility for their lives and provide fantastic teamwork. Why? Because we can only really achieve success if we help others to be successful! Regardless of the level that we have reached, we can only make progress if we help our partners in the team to achieve the same! There is no other way.

This is how we achieve our "spirit," our so-called "USP" ("Unique Selling Position"), i.e. our singularly outstanding feature! And I like this. And the people who really want this feel comfortable with us and are not under any pressure whatsoever. But what they do have is a real stable income that is constantly increasing (albeit slowly to begin with, I confess)!

And the products? They are simply top class. So good that people are happy to buy and use them. That goes without saying. And they are important, of course – extremely important! However, I have found that many people have not yet thought about this subject and therefore they are simply not interested in it. Or they have been frightened off by over-zealous"sales people" and have shut the product door for entering the farm.

In spite of this, I would like to offer all of these people the opportunity to change their lives and therefore open the other door: the door to visions and dreams. In my experience, this is the door that the majority of people go through. At the end of the day, it is only a question of the order in which we do things.

I personally appeal to **every** new partner to get the right reading material so as to have peace of mind that good supplements are useful and actually indispensable for the body. My experience has been that those who have started for other reasons, sooner or later, what nutritional supplements can do for them. And this is my goal eventually! Another advantage is that a product user will keep buying the product even though he may not want to do the business.

I now know how confidence grows if the new partner himself, or a person in his circle of friends and relatives, has experienced the effects. Suddenly, this person has a completely different gut feeling! They speak with enthusiasm and with inner conviction, and have a completely different charisma – and suddenly they are successful! It is also because of this that we need to perform "personal market research" within our family. This doesn't mean that you can't talk to people until they have achieved this gut feeling. Your tools and your sponsor are there for you at this early stage in the proceedings, after all!

I would like to explain to you why this chapter is so important by telling you a story:

A female partner was discussing pre-information and catalogue information with an Italian lady whose husband had passed away several months earlier. After his death, she and her daughter discovered that their financial situation was difficult and they were looking for an alternative in order to survive. They were immediately enthusiastic about the opportunity that was offered by Referral Marketing. When they were introduced to the company's products in the catalogue information, the mother started to cry. She had already used the product with extremely good results, but had to cancel the order after the death of her husband because of their financial problems. Her sponsor had not told her about the possibility of refinancing her product, or about the opportunity of building up a second income, or even a business of her own!

She knew nothing about the farm. Who has the right to judge who needs an opportunity and who does not? I cannot do it, so I always provide everyone with all of the information.

I'll just summarize it again. Of course, we need customers and product users. But I let the other person decide for themselves whether they would like the chicken, the egg or the farm. This means that if someone does not want to be actively engaged in Referral Marketing, I still offer them the products.

Do you get my drift?

The name list

I must admit that I have changed my mind completely about the subject of "name lists." In many Networking books you read that you have to create a "list of names" or "list of people" or "100 list" when you first start out. I always felt a little uneasy about this, because I simply talked to everyone about my opportunity without reservation. I immediately saw the opportunity, and nothing could have stopped me. In my previous days of direct selling, quotas of 20 contacts per day were nothing unusual. Advertisements were in place all the time, which meant that finding five people who wanted to take part was a "joke" as far as I was concerned. It was the same for Lissy and Isolde, who did not need a list either. However, I underestimated the power of this list. Thank God that we can change our minds.

I changed my mind completely and now think that this list can be the basis for survival to some people. There is a special reason for this:

Everyone knows at least 200 people in their immediate, close and remote circle of acquaintances. 100 of them suffice for the list to start with. Hopefully, the new partner will have influence on several people in this list and is considered to be credible. It will therefore not be a problem to ask a few family members and friends for help. We need a few product users for "personal market research," so to speak. With our relatives we don't need to be "good" and "trained" – it suffices for us just to be ourselves and say:

"Aunt Annie, you know my situation. My company is about to fold and I don't want to wait until I lose my job. I have just started my own little business on the side and you could help me. Will you do that?

And so on:
My company has wonderful natural products. It would help me a lot if you ordered the products you need anyway from my company. This is the catalogue…" They have also fantastic products for prevention and health. I will give you an excerpt of a very good book by Anne Simons. You can read it and then we can talk about it later."

Few people would refuse this request for assistance. And this produces results that are extremely important for our own security, our gut instinct and our confidence in the possibilities of Referral Marketing.

It is not so much a matter of "Selling and Convincing" as "**Filtering and Sorting.**" This is important, and is all about getting together with our new partner when they start out and finding four or five people in the list of names who are looking for a way of changing their situation **here and now**.

The "quota" also plays a part, i.e. the number of people I have to talk to until I sponsor one of them. Not everyone who is on your list will be immediately enthusiastic about setting up a business or trying out a product (not yet?). This is simply the truth that we have to live with.

Suppose your quota was **10:4:2:1**. This means: if we introduce the opportunity **ten** times, **4** people order on average, **2** just use the products, **2** actively recommend the products and of the two, **1** of them starts up properly and **1** only gives random recommendations. As a professional Networker I now look at the subject completely dispassionately. I see the sponsoring of new partners like a lucky bag – I never know what is going to come out of it, and surprises are always possible.

I probably don't need to tell you that the quota improves together with the frequency with which conversations take place. Here too, we benefit from experience.

The quota differs entirely from person to person, and depends on factors such as positive attitude, credibility, ability to communicate and the relationship with the discussion partner. But one thing is certain: it is proof that anyone can do it!

Someone with a low quota simply needs to carry out a few more conversations! But the great thing about it is: We have time, and ultimately the message is:

"Perserverance beats speed!"

I think that if a new partner is aware of this from the outset and accepts this as **NORMAL**, there will be no problem if people say NO. The partner then knows that this is important for success. Because in a similar way to my example, he knows exactly that he needs six "NOs" in order to get four "YES" answers.

This is why the length of the list is so important for psychological reasons. Imagine that a new partner only has five names on his list, and three of them say "No." This would be a catastrophe.

David Ledoux describes this topic in great detail in his book "30 Most Dangerous Documents." He considers it extremely important to draw up this list together with the new partner. Because otherwise it would never exist… Presumably he is right, we will have to draw up the list together with the majority of new starters.

How do I draw up a list of names in practice? Here are a few tips: Take the contact form from your sponsor or the member area, or simply take a pad and pencil and write down: Close friends, relatives, acquaintances, neighbors….

Then consider the fact that every person who you record in your list could be someone with whom you can build up a business and make yourself financially independent. Do not make judgements. This means that you do not decide who it may be suitable for and who not. You cannot tell from looking at someone whether they need an opportunity or not, so we should therefore tell each person our story and therefore give them a chance. When this small group has been drawn up, the list can be extended as follows:

- Daily routine: baker, hairdresser, travel agent, doctors, non-medical practitioners, car dealers, cleaners, postman, mechanic, landlord, artist, insurance representative…
- Professional: former and current work colleagues, trade or business partners, former and current customers, former employers, freelancers, travel acquaintances, trade fairs, tax advisers, bankers…
- School/education: kindergarten, school friends, adult education, fellow students, teachers, seminar participants, evening class…
- Sport/Hobby: fitness club, dance club, painting course, swimming club, golf club…
- Club/Association: parents committee, political party, parish…

The important criterion for your conversation is to find out why the person could be interested in looking into your business opportunity in more depth. I have already dealt with the **WHY** in the first part of the book.

The more disciplined you are in drawing up the list of names and collecting facts for these people, the better your chances of success.

Note down everything that you know about the contacts. "Hobbies," "Profession" etc. Note down all details such as "built a house," "installments to pay," – "wants to buy a car," "likes to travel around the world," "single parent," "likes fishing," "likes sailing" or whatever else comes out in the conversation. Be **attentive**. Attentiveness is extraordinary and has become extremely rare in our world. **Attentiveness is noticed!**

And then make contact. If you have already spoken to someone and they are in the "Not-Yet-Box," **maintain the contact**. This may be in the form of a congratulation on the event of a birthday or an anniversary, a Christmas card, an invitation to a personality seminar or your garden party, or calling to inquire about their health. If you know their hobbies, you can send a newspaper article about their hobby or another topic that the person concerned is interested in. Or what about a reminder about a television program on the subject of their hobby?

If you know that one of your contacts is a passionate stamp collector, simply send a postcard from all the places that you go to with a nice stamp on it! Can you imagine how pleased that person would be?

A basic thing to remember is:

When you contact a person again, do talk about any subject, but do NOT talk about Referral Marketing!

Ledoux says in his book: *"Communicate until the person signs up, dies or moves away…." By the way in our case moving is not an obstacle. It is an opportunity because we work in 64 countries.* The list of names will change, of course. The aim should be to add new names every week. A list like this is never empty. I have not even started mine. But as soon as this book is published I really will make a start. This could also be a "pitch" for people who have been in the business for some time. A tip for the "old hands."

If you enjoy my book, take your list of names and contact everyone in the "Not-Yet-Box" and tell them about it!

The "Not-Yet-Box"

I have told my story to everyone who has asked me to. And the majority of them have asked. Why? Well, because I asked **THEM** beforehand what **THEIR** profession was… That is the difference. I can wait until someone asks me by chance, but I can also reach out to people and show that I am really interested in them! Some people have asked, but then turned out to be not interested. That is a matter of the quota. But due to my experience one thing I know for sure:

There is no such thing as "No" – there is only "not yet." When I notice that my conversation partner is not interested in further details after I have told my story, I do not take this reaction personally. Because this has nothing to do with me, it is merely a matter of the timing. I simply put this person into my mental "Not-Yet-Box." Perhaps the timing wasn't right. I have sown a seed and all that I can do now is nurture it. That is the major advantage of working in a warm market.

"Nurturing" means working **actively** on the **list of names**, as I mentioned in the previous chapter, and carrying out "**unintentional relationship development.**"

There is also another type of "Not-Yet-Box." This concerns people who want to become active but have **NOT YET** started. For these people I also have an appropriate story:

In the year 2000 I sponsored a couple from Bremen on a tour of northern Germany. On the way back we stayed over in Brunswick in our camper and visited Lothar, who was a former colleague of mine. Lothar ordered the products but had asked a friend for advice with regard to the business opportunity, who told him to leave it be. Fortunately, the product worked so well that he continued to buy it. A year later, the couple from Bremen had reached the highest level of the compensation plan and I called Lothar again: *"I currently have time to support someone again. Do you want to start or wait for another year?"* He started after his friend Jörg had also been persuaded to sign up. Who then sponsored Uwe. And do you know where Uwe now lives?

On Majorca!

Can you imagine what that meant to me? We immediately planned a business trip to the island.

Uwe already had a considerable amount of Networking experience and carried out excellent set-up work, meaning that 36 partners were present at the first seminar on 29th June 2002. What was the result?

The islanders turned out to be a great team. Eleven of them have now reached "Diamond" status, three of whom achieved it on the island. They are now generating a total of about 8 million Euros per year! So do not underestimate the "Lothars" who just use the product for a long time.

I spend a considerable amount of time on the island. It is simply fantastic there, and to me it represents pure quality of life. The light is fantastic. We go for walks in the mountains, usually with a view of the sea. Like I said: who says that training has to be carried out in a room? The training sessions in our pool have now become famous, where we are literally "up to our necks in it" during our work in the literal sense of the word. I now have more friends on the island that I could ever have imagined.

I love Referral Marketing!

If you understand this, you will know why I advise you to sponsor people that you really like....

I simply offered Lothar the opportunity at the wrong time. Or he didn't trust himself to do it. I have several people in my group who need more time and use our seminars to develop their personalities. That is one of the issues that lies close to my heart.

I recently received a Mail from Annelie:

> "You asked me about my goals. I struggled with this question because I feel trapped. I would love to stand on the stage like you and tell people about my experience with this great Referral Marketing. I have read many books about Networking and have enjoyed every one of them. But I have a problem with the implementation. I simply feel paralysed and ashamed that I can't do it. I hope that I will not be dropped and that I will be given time to get over it. I heard on a CD that Tom Schreiter took two years to sponsor his first partner, which was a source of some comfort. However, I don't want to take so long."

My words to Annelie and all the other Annelies in the world are: take all the time that you need – you have all the time in the world! Your story could have been: *"I have discovered a way of making myself financially independent. At the moment I don't have the confidence to do it and still haven't started. But just having the opportunity to develop my personality and my capabilities is fantastic!"* That would be honest – and I can imagine that someone or other would ask you about it. And whoops! – you are having a conversation without any pressure! I would just like you to do one thing: let your sponsor known that you are in the "Not-Yet-Box."

In our business everyone is their own boss, without turnover targets or pressure. For this reason it is particularly important for the impetus to come from you. The onus is on you to take responsibility.

Some people who we talk to may have had a bad experience with direct selling and don't want anything to do with it at the moment. Believe me, that is no problem. Situations change. Think about my brother and Lothar.

A **NO** is never final, because your attitude can change tremendously in many respects during the course of time as your situation changes. Many people don't see how it is developing as far as they are concerned! Nowadays I would say:

"You can sponsor anyone if you take enough time and build up a relationship. All you have to do then is be there..."

Nearly all of Lissy and Werner's family are now taking part. Her sister joined after a year, and her brother joined after four and a half years. With Werner it was somewhat sooner. One sister (Elfi) in Switzerland joined after six months, and Sigrid and brother-in-law Walter followed three and a half years later. His brother Dieter was a product customer right from the start, but did not actively become involved in the business for four and a half years. When Walter received the Bronze award on stage, he said: *"I watched the whole thing from Switzerland and am extremely grateful to my brother-in-law for never pushing me!"*

We want to use the network, not stifle it!

I therefore also recommend:

Inoculation and snail technique

I discovered some important tips in a book written by Mark and Rene Yarnells that I mentioned earlier, which led me to this matter.

They write that 95 % of all Networkers who stick with it achieve their goals. Unfortunately, the number of Networkers who don't get this far is extremely high. In order to prevent this, the Yarnells have a passage in their book that amused me no end:

"I know that some of you are thrilled with the numbers that you have heard today. It is also clear to me that you will want to think about the whole thing carefully again, like every responsibly thinking company. But let's look the facts in the face: if you really can achieve this monthly income in order to be half way to retirement in three to four years or at least have considerable personal and material freedom, there must be something wrong with you if you cannot work up any enthusiasm for my suggestion. However, at this point I would like to warn you about the two main reasons why you could fail. So that you can avoid them during your subsequent research: the first reason for failure is when new Networkers

listen to people who do not know what they are talking about. The second reason is not listening to those of us who do know what they are talking about...."

I had to laugh, but there is some truth in it, and I'll leave it up to you as to whether you use the "something wrong with you" bit. What is important to me is that I prepare my partner for the fact there are people with prejudices after we have had a conversation.

Inoculation

> **Do not listen to people who do not know what they are talking about, but listen to those who have experience in this area and have been successful.**

The opportunity that you are being presented with by Referral Marketing for shaping your future is simply too precious. Do both of us a favour and only listen to people who have already achieved what you want to achieve too. Someone who has not been successful cannot give good advice on how to be successful. Take the time to gather information and convince yourself that this is a respectable line of work and a respectable company. Don't be deprived of major opportunities because of minor issues!

I would now like to introduce you to what I consider to be the Networker's most important tool. Without which he cannot operate. It is therefore unbelievably precious, and every Networker must ensure that it is not stolen from them. What do you think that this instrument could be?

Correct – it is our **energy**, which is unfortunately taken away from us sometimes by unnecessary, fruitless conversations. And then we are drained for three days and don't trust ourselves to have any further conversations.

The snail technique

The snail technique must only be used in the early stages and only has one purpose: to avoid the arrow of rejection. The snail withdraws into its shell if it meets an obstacle. This helps new Networkers to retain their energy. Once you have been doing it for a year you will not need it any longer.

Do you think that your future should depend on the opinion of the first person to whom you told your story? One person is enthusiastic because the friend to whom he told his story was just looking for a way to get away from his boss and started immediately.

At the same time, another person quits the "career" that he has just embarked on because the first person that he met had a garage full of products several years ago and doesn't want to hear anything about "snowball systems!" One person tipped the scales! For this reason, an important request:

Don't try to discuss the matter if you receive any objections, but say:

"This is interesting. Thanks for pointing that out. I' ll look into it."

And then withdraw into your shell. This means, go to your sponsor and have him explain it to you. If you sponsor is no longer active or has emigrated or even died, go to the next person in your Upline. Look for a mentor! If you do not find one, just call me. I will find a person in your active Upline because I do not like to leave someone in my team helpless.

Successful listening

Learn to be a good listener! There is a small book with this title in our book list. The longer I am in the business, the more I realize how important it is to be able to do this. I admit that I do not find it easy and I am still somewhat challenged in this area.... Basically we only have to wait for a keyword from our partner in order to butt in at a suitable point and tell our story. Honestly, how often have I heard comments like the following?

- *"I am fed up of never having any money. I would like to take a holiday again some time."*
- *"We hardly see each other because we work so much. My husband and I do nothing but come and go in a continuous flow."*
- *"Our son wants to study, but we don't know how we are going to pay for it."*

- *"My job is on the line. They want to close the factory."*
- *"I don't just want to be the little woman for the rest of my life!"*
- *"The children have grown up and I want to do something and be among people again."*
- *"I' ll never get my old job back after having children."*
- *"I found out last week how much my pension is going to be. I nearly fell over."*

These statements are ideal for telling our story:

"Tell me about it! I don't need to worry about my pension any more because I have just discovered a brilliant way to build up a reliable pension income on the side."

However, you will also be able to offer someone or other an opportunity for the future with this business proposition by asking questions as well as listening. Because questions are the answers. Ask you conversation partner and then listen. People like to talk about their feelings and their dreams. We are all glad when someone "lends us their ear." You can help other people to make decisions that affect their future by listening properly.

"Listening For Success" by Shapiro contains an interesting paragraph:

"Listening is like picking fruit from a tree (...) If you interrupt someone, finish their sentence, talk too much or reply too soon, it is as though you are shaking a tree before it has had the chance to grow any fruit....

Let us suppose that you are listening to someone telling you about their dream of working from home. The majority of Networkers would now start to tell this person everything about their fabulous business and would talk, talk, talk. A better reaction, and one that would bear fruit, would be to listen:

"Have you always dreamt about working from home? Can you tell me more about it?" or: "What do you find so attractive about it?"

Asking questions and listening allows the fruit to ripen on the branch. And fall to the ground of its own accord before your very eyes."

This could be the reason for my success. I am extremely decent and patient. And I let the fruit ripen on your tree and never "propose on the first date."

For this reason I am now convinced that we can win over almost anybody for our business if we find out their **WHY** and offer a solution. And if we are patient. The opportunity is simply too good not to be taken advantage of. And why am I so certain? Simply because it has major advantages and no disadvantages! Taking all of the "objective facts" into consideration, even the greatest sceptic would come to the same conclusion in my opinion....

I am certain that most people know too little about our business and the industry, and we know too little about them. You will find out the **WHY** sooner or later by listening, and will build up a real relationship that will then be the basis for trust.

Contacts

In my opinion, the procedure should be divided into two completely **different** areas:

Firstly, people who we know and have been introduced to us by our friends, and secondly: people that we do not yet know.

Basically it is our job to find five of our best friends and help them to achieve their goals by helping them to build up a team. In this sentence it is clear that if we really do operate our business in this way, we are in an absolute "**offer** position," and not an "**ask** position" under any circumstances. There is a major difference between the two! The relationship with these people has already been established, which is a major advantage!

The Yarnells write in their book "Power Multi-Level marketing":

"(...) this first phase of working with friends, family and acquaintances is an important step in gaining the experience and the working base you need to move on."

And *"We have to tell you, however, that if you are one of those people who is reluctant to contact friends, relatives, neighbours and acquaintances because you are afraid of leaving them with a bad taste or impression, then you are in the wrong business. We do have some suggestions, that might help you overcome your reluctance. First, do not deceive people. (...) Tell them you have a great money-making opportunity. Or you might indicate you want their advice about a new venture. (…) You should also know that you are really doing your friend or relative a favour. Remember that in recent years, a significant number of new millionaires have come from this industry."*

Or, as Don Failla said, *"If you are really convinced that you could retire in one to three years, why not offer the opportunity to a stranger instead of introducing it to your friends and relatives first?"*

And now I'll tell you the big secret!

If you **really** help five active people in your circle of acquaintances to build up their teams, you will never have to concentrate on how to make "cold" contacts. I never needed a list of names. I always had enough to do by helping my people to sponsor and train their own partners. That is the perfect training, and provides independent partners – without whom Referral Marketing would certainly never be able to lead to independence.

As Don Failla writes: *"You will never have to speak to strangers again. Meet a friend and talk to their friends!"*

Unfortunately, this principle is neglected in practice. Perhaps it is not fully understood, and I would therefore like to make the following appeal, particularly to the old hands:

Are all of your partners in the team happy with their situation? If not, they still have growth potential. With their existing people. Without having to make new contacts. Of course, we need determined partners in order to do this.

Partners who are committed to passing on the assistance that I have given them in the same way: *"I promise to provide you with a regular income, and in return you promise to provide your Downline with a regular income."* This is how teamwork works: a constant flow from top to bottom. Stop wanting to receive, and start to give.

But to be perfectly honest, I have never received a "**NO**" when I have asked my existing partners:

"I need a new star. Which one of you will allow me to make you successful?" (Our highest status is "Diamond." And for each of my lines that I help to become a Diamond I receive a star. At the time of publication I have six). Here too you can see that we can only become successful if we help a team or a person to become successful. Simply generating turnover is useless to me.

Referral Marketing works in places where life is going on, between people. I would advise you to spend time in the places where your friends meet their friends.

And where is this? At the morning jog, for example! If you don't have one, then start one up! Lissy and I had our first conversation during the morning jog and invited our potential partners to it as well. A three-way conversation such as this has influenced many a decision.

You can also invite your sponsor to your birthday party. And then introduce him/her. *"This is Susanne, I am just building up a second income for myself and she is helping me!"* Of course, a birthday party isn't ideal for going into detail. It is just a matter of letting your sponsor make contact with your acquaintances and building up a relationship. Personally I like to use the measuring tape story – it is always well received and I tell them that I have achieved my goals. If they ask questions I say: *"We're here to celebrate a birthday, but I am here on Friday for a training session – maybe it would be best if you popped in then to see what you think!"*

If your birthday is a long way off, invite them to a barbecue or dinner. Like I have said, it's not about business – it's just about building up a relationship!

I say this deliberately, because there are also so-called "cover-up meetings" at which a supposedly resourceful Networker invites his friends to dinner and then makes a business proposition. I do not like these methods and think that they have already done enough damage to the reputation of our profession. We don't need it!

Speak to your friends. Praise them: *"I thought about you because you have always got on well with people!" Or "because you are so understanding!" or "because you have always given me the impression that you understand something about good business."*

By the way, praise always works – with **EVERYONE**. It also has to be honest, of course. I could tell you stories…. The praise that I have received is all in a file and I can still remember the names….

If you are completely new, you start by looking for positive things in the people in your environment. You will find some. And then say something about them! Your appreciation level within your circle of friends will double within six months! I can highly recommend the book "How to Win Friends and Influence People" by Dale Carnegie to anyone in this respect.

I recently read a book entitled "Erfolgsfaktor Networking" by Uwe Scheler ("The Success Factor of Networking"). This is a book about networking in your private life. His advice is extremely useful and generally applicable. For example, I read *"Institutional and personal networks are a system of mutual giving and taking…,"* and *"Fortunate circumstances are one thing, but recognizing and exploiting them is something completely different. Successful people differ from those who are less successful by being able to recognise favorable circumstances and exploiting them."* Under the heading of "What to avoid" I found *"Never attempt to sell anything or offer your services to anyone at a cocktail party. You can discuss your offer in a relaxed way, but never carry out sales discussions."*

If your new partner likes bowling, simply go along. Perhaps your new partner's rabbit club has a garden party at which you can harmlessly show your face or…, or…, or… See – **there are hundreds of possibilities! Be creative**!

Now to the second area. Back to contacts of people who have asked **all** of their acquaintances and **ALL** have said "**NO.**" (I have to smile as I write this, because I cannot believe it). And people who have just moved and do not know **ANYONE**.

In general, I recommend spending time in places where active, open-minded people are present. Seminars about any possible topic are suitable, for example. Let's take an adult education course in English or Rhetoric as an example. These are both topics that will be useful to us. And we will also meet people. People who are open and thirsty for knowledge (otherwise they wouldn't be there!), which is a major advantage!

One of my aims was to have enough money and time to develop my personality. There are so many seminars that I still like to go to, and the way in which they are organized is such that each person introduces himself and tells their story. I have never failed to tell my story without several of the participants subsequently asking me about it. And I have never been to a seminar that failed to generate a new partner.

Inviting third parties

As you already know from the introductory story, in July 2004 I was in Switzerland, where I carried out several training sessions. An event of this nature can be used by anyone in my group as a "pitch" (opportunity to speak to someone in a neutral environment). In this case it might be a reason to get out the list of names and see whether someone is there in Switzerland – or whether someone on the list knows someone in Switzerland.... I can then call and say *"Hi Egon, I was thinking about you. I am now selfemployed and my company is expanding in Switzerland. Next week we are holding an information event. Do you know someone who wants a change of profession or is looking for a second income?"* It is possible that he might say *"Yes, what's it about? I might be interested myself!"*

In any case, this could result in a conversation and I have the opportunity to make my offer.

In passing

Never try to introduce Referral Marketing or your company in passing. *"He would obtain enough information to refuse, but not enough to say* ***'YES',*** as Don Failla writes in his book. After you have aroused curiosity with your story, it is better to say: *"The opportunity is too valuable to mention in passing. I' ll explain it all over a cappuccino. When do you have time?"*

The one thing you can certainly do is give them an independent tool beforehand. My contact must not have the impression that I do not **want** to say anything – that would make them mistrustful.

The "reverse logic" or the "you don't need it method."

People have a tendency to react in unexpected ways. Basically, we know that there are many things that are normal in everyday life that we like to forget when we start out in Referral Marketing.

There is one example that I am particularly fond of: When I say to my friend: *"Hey, I have discovered a great anti-aging cream that would be good for your wrinkles,"* she will probably indignantly say: *"What are you talking about – I don't have any wrinkles!"* But on the other hand, if I say *"Hey, I have discovered a new anti-aging method that prevents wrinkles, but* ***you*** *don't need it because you don't have any wrinkles...,"* then she will probably pinch her cheek with her thumb and her index finger and say: *"You are wrong – look at these wrinkles!"* Do you understand what I mean? Just try saying to someone you know: *"I have started to build up a second income, but you have no need to because you already earn enough..."* or *"You already have enough money."* It works (nearly) every time....

The delayed approach

In her book "Owning yourself," Paula Pritchard describes how to awaken people's interest. She notes down the points that occur to her using the conversation and then calls later: *"On Sunday you said that you wanted to own your own house. Were you just kidding, or were you serious?"*

Of course, nobody says that they were just kidding, and she can then present the opportunity to them.

This "delayed" approach is described in several books and I find it extremely good, particularly with friends and acquaintances that I will definitely be seeing again. Remember what I said: With my closest circle of friends and acquaintances I have the opportunity to sow the seed and then nurture it.

This means that I do not necessarily talk about the business during the initial conversation. I would like to avoid the impression that I am only interested in a friendship because I would like to sponsor this person. If someone comes over as being sympathetic, I definitely want to get to know them. And sponsor them, of course, because I like spending time with them and would like to help them. I don't have a problem with this because my attitude is simply that I have the best opportunity in the world to offer them.

As I said previously, in my opinion you can sponsor anyone if you invest time in people and build up a relationship.

Patience is an important factor in our industry. I always like to tell a story that I heard years ago about the bamboo tree. You plant it and water it. You wait for a year, and there is nothing to see. Then you water it for another year –without any signs of success.

In the third year you probably just water it out of habit and in the fourth year the tree suddenly grows to more than 20 metres in height! Now the question is: how long did it take? One year or four years?

The situation is similar in Referral Marketing. Of course, I don't want to give the impression that you can plant a seed and wait until it grows of its own accord. Or even sit on it and hatch it without doing anything (unfortunately a form of Networking that is often used). The secret in Referral Marketing is that we simply continue to sow and nurture. Preferably without worrying about the harvest. And without spending too much time on people who do **not yet** want to.

This is why we need to have some 15-20 active partners in order to find five serious ones. And I should probably also stress that we do not need all five at the same time! I think that anyone can achieve this with the help of a sponsor.

Our website

For a long time I thought it was unnecessary to have a website. Today, however, I see it differently, at least when it comes to the purpose a website should fulfil! I still believe, however, that it is **not** possible to build up a successful network simply on the internet, without personal contacts.

Years ago in "Wave 4" Network-Marketing in the 21st Century by Richard Poe, I found a chapter which I have never forgotten:

"The 3rd Wave mania is threatening to turn MLM into a business without relationships. This could harm networkers in many areas. The use of automated advertising methods such as emails, websites and faxes sent at random lead to people having downlines of 10,000 people from whom you never hear again. This is not the way to build up a residual income. The simpler you keep Network Marketing, the better it works – by talking to people every day, one-on-one and from the heart. Friendship, loyalty and personal relationships are the glue that keeps a network together over a long period of time."

I can only support this! If we only relay information via email and do not follow up with a call or personal visit our team will starve emotionally. We may, however, only find that out later…or perhaps never. You can never count the many business opportunities you may have missed!

A website can be a beneficial tool for giving friends, family or other contacts a comprehensive overview. In the meantime, we can have our own website which we can then personalize. In this personal website, we will introduce ourselves with our own story and explain our business in a few brief sentences. Independent books, articles and tools are also used together with neutral "pitches". The most important of these are the "success stories" and references from highly regarded lawyers, doctors and other opion leaders who endorse the quality and standing of our business.

Information flow

Referral Marketing works by using communication of information and promotion.

During all of my training sessions I often ask the participants to guess what the most important capability is in Referral Marketing. I now know from experience that someone can be successful in spite of having **ZERO** capability if they succeed in moving people "from A to B." This capability of promoting puts everything else in the shade….

The film "Paying It Forward" was recently shown on television, and depicted an excellent example of how threefold duplication networking operates. The story is about a young boy who wants to improve the world. He does a favour for three people, provided that they do a good deed for three other people, i.e. "paying it forward." These people would then do something for the next three people and so on. I had read the book "The Wonder of Innocence" and when I saw that the film was going to be shown on television I promoted it. (The book was much better). Still, the film was a great "pitch" – i.e. a reason for promotion. I would like to show you the four levels of understanding in Referral Marketing using the following example:

When I call my Anne and she says *"Yes, I'll watch the film,"* then she has understood the first stage. If my Anne now says *"Yes, I will watch the film and tell my other partners to watch it as well,"* she has understood the second stage. The third stage would be: *"Yes, I will watch the film, tell my partners so that they will also watch it, and I will also tell them to promote the film to their own partners."*

Once this stage has been reached, the circulation is perfect, right down to the final toe! And we reach the highest level of all when we also consider telling acquaintances who have been on our list of names for some time but have not sponsored about the film. Then we tell our Annes to tell their Annes to also consider the people on their own lists of names.

When I discovered this system in 1999 I became aware of the dynamics and power behind it when the "circulation works right down to the final toe." This means that the person who is furthest away from you (measured in levels) also receives the information! Imagine that you have been using the system for a year and have 780 people in your Downline. Something new now comes out (perhaps this book?), or there is a seminar to promote. Providing that there aren't any "circulation problems" it would operate as follows:

I send an e-mail with the date. Then I immediately go to the phone and call my five Annes. *"Hello, I have just sent you an e-mail – Jörg Löhr is coming to Stuttgart. That is superb – we have to go. Please call all of your Firstlines and promote the event. And tell them to pass the date on to their own Firstlines!"*

That would cost me five phone calls. When my Annes have told their Bernards, 25 people already know about it, and if they call their Christines, 125 people will know about it. Just think what would happen if a network operated efficiently. How long would it take me to make the calls? One hour, perhaps two?

This means that if I notify my five Annes at 9 am, 25 people could already know by 11 am, 125 would know by 1 pm, and 625 by 3 pm. Our entire Downline with 780 people would now have received the information! Do you see what I mean? And I only spoke to my five Annes.

Unfortunately things are different in reality. I think that the majority of people do not understand these dynamics, and are therefore (still) not using the system correctly. What happens if only four of my five Annes pass on the information? What do you think we would end up with instead of 780 people? Normally it is 80 % of the total – at least, that is what most people would estimate. But this is just additional evidence of the fact that the power of Referral Marketing is not easy to understand.

We would end up with just 425 people rather than 780! This means that almost half of your Downline would not be notified!

Duplication also works in the negative sense!

There is zero tolerance as far as this point is concerned: If you introduce a new partner to the system, ensure that the partner is hooked into the flow of information (www.mitgliederbereich.com) and then call them anyway and **promote** that which is important. In my experience, only about 20 % of e-mails will be read. (I am quite sneaky and keep carrying out little tests). A call is answered up to 100 % of the time. Want to increase your quota five times over?

Then GRAB THE PHONE…!!!!!!

Typical questions

Of course, questions and objections crop up all the time during conversations with interested parties. The same issues come up time and time again, making them easy to deal with. The most important point when doing this is to find out right from the start whether someone just wants to say "**No**" and now has all sorts of objections and questions, or whether there is any real interest there. Being able to tell the difference is useful, because spending a lot of time on the "first category" is nonsensical and wastes a lot of energy.

On the other hand, the second category is extremely welcome. I always train my team to trust their gut instinct in order to be convincing.

Of course, objections that are more or less typical also occur time and time again. For this reason I have put together a selection of objections:

"I'm not interested."

The Yarnells have a good reply to this: *"I can understand that. The job isn't everyone's cup of tea. But in order for me to understand you better, precisely what is it that you are not interested in? More money, Or more freedom?"* (That is a jest, of course – my conversation partner is put into the "Not-Yet- Box").

"My other half objects."

In my experience, this happens quite frequently – above all if a partner has been informed alone. I therefore strongly suggest carrying out the conversation in the presence of the interested party's partner. And if this is not possible, tell them Lissy and Werner's story!

"I know about it already."

My reply to this (and I mean it seriously) is: *"I do not believe that you know about it already. Because if that was the case you would go along with it!"* Everyone has already heard about Shell and Esso, but do they know all the details? Just ask what they know already.

To be honest, I think that it is a big advantage if we are the second person to whom someone speaks about our business. Haven't you ever been spoken to about the same topic twice in quick succession, and thought: *"I'm hearing it now for the second time within a short period, it's about time I found something out about it!"* At lunch today I spoke to Tommy, a friend of many years, who started in the business this year. Yesterday he was the second person who was spoken to by someone, and discovered that although the facts were interesting the first time around, the chemistry was not right! What he said was: *"I am happy when I meet someone who has already heard about it. This means that someone else has already done the groundwork!"*

"I have a healthy diet, therefore I don't need vitamins."

I used to discuss this point. Now I stick to Lissy's advice and say: *"I find it amazing that you manage to eat five to seven portions of fresh fruit and vegetables every day! Well done! Being a businesswoman I can't manage to do that any more. And the health food shop isn't exactly cheap…!"*

That is fantastic! The reactions differ greatly, because I doubt whether **anyone** eats even three portions of fruit and raw vegetables every day. (The statistics say that the proportion is 1.2 portions per European, by the way).

We are all well aware that cooking takes out the vitamins. And who has the time to go to the market every day?

But there is another completely different reason. This topic fascinates me and is close to my heart, for which reason I would like to give you some stimuli. I recently got my hands on a new brochure about OPC. Brief, concise and inexpensive – to put it simply: fantastic! It describes a topic that I find quite special:

"Most scientists are now convinced that a person can live for more than 120 years. When you reach the age of 60 there is no reason why you shouldn't live for another half century or so!"

The brochure also says:

"The figures from the WHO (World Health Organization) are ancient and the way in which the requirement figures were worked out in those days sounds like a bad joke: test subjects who volunteered were denied a vitamin until the initial deficiency symptoms occurred. Then they were given the vitamin again until the symptoms disappeared again. The dosage that was required represents the daily requirement according to the WHO. I can assure you that taking 100 milligrams of vitamin C every day may not make your teeth fall out, but a harmless cold virus will probably laugh its head off at your attempts."

The possibility of living for 120 years is not an individual opinion. Because of my many years of intensive study of this subject, I am convinced that it is possible. And I am trying to achieve it. When I talk about it I sometimes receive the response *"I don't want to reach that age!"*

When I hear this, I am the one that is laughing – I haven't met anyone who, as Strachowitz says, *"makes use of their right to a socially acceptable early demise."* I provide the appropriate response: *"It depends on how much pension you will get. With my income it's worth living until I am 120!"*

Of course, there is more to this than a healthy diet. There are plenty of books that will tell you about it and that you can read if you are interested in the subject. As well as top quality food supplements (and don´t be stingy like the Swabians), this includes exercise, plenty of water and a positive attitude to life.

The most important thing of all nowadays is the need to detoxify our bodies. Vitamins can take over this additional function. Whilst we are on this subject I would like to tell you a story from a book by Dr. Strunz, the wellknown "Fitness Guru."

"Do you know the story about the chicken and the nickel?

Some researchers divided some chickens up into two groups. One group was given quality food, and the others were fed cheap stuff (fast food for chickens). Then the researchers mixed nickel with the food that the two groups were being given. Of course, we know that jewelry containing nickel causes allergies on the ears and stomach and is fairly poisonous on the whole. After a few weeks the chickens were slaughtered and examined in the laboratory. And what did the researchers discover? There were no traces of nickel in the chickens that had been fed with the quality food! There were only slight traces in the liver and the kidneys, which are the detoxification organs. The group that had been given the 'fast food for chickens' had nickel in their flesh and the liver and kidneys were riddled with nickel."

Dear reader, you are currently taking part in this experiment!

...and which group are you currently in?

"Too expensive"

This is a topic that usually emerges with people who are still "selling products" and have not yet understood how Referral Marketing is different. Who can tell the value of a product at a single glance? What are you comparing it with? One of our main products is OPC (Oligomer Proanthocyanidin – you'll soon have forgotten what it means). This substance was discovered by Professor Masqualier, and Anne Simons wrote a book about it called "Living healthier for longer with OPC," which you can buy from any bookstore. There is now a work book about it containing a questionnaire and 33 questions: "Do you need OPC?"

There are more than a million web sites about OPC worldwide, which has also been labelled the anti-aging vitamin of the millennium. An influential beauty magazine wrote that OPC is well on the way to establishing itself on the market as an anti-aging substance. I would like to use this example to show you how difficult it is to compare prices. If you encounter a product known as "Grapeseed extract" or "Pycnogenol" you are dealing with a simple grapeseed product. "**Proanthenol**" is much more than just OPC! **Proanthenol** is a mixture of grapeseed extract and pine tree bark and other ingredients that have a high synergistic effect (**1+1=5!**).

OPC reinforces the effect of all antioxidants (such as Vitamin E and Vitamin C) several times over. Many people have Vitamin C deficiency. For this reason, the company that we work with has enriched the **Proanthenol** with an additional 200 mg of Vitamin C, **querticin**, **rutin** and **hesperidin**. This extended OPC is 100 % **biologically available**. The high proportion of cellulose, which most companies use as binding agent, has been replaced with the synergistic **phytozyme basis**. Proanthenol is therefore a phenomenal further development of conventional OPC.

This topic is extremely interesting. Many companies are somewhat careless with this matter. The proportion of pure OPC and above all the **biological availability** are also extremely important. Tests have proven that there are differences of the magnitude of between **5 % and 100 %**. How is a layman expected to be able to assess what he is paying his money for?

Real, top quality OPC has its price on the global market. A gold ring can never cost less than the gold from which it is made. The price is therefore a yardstick of quality. And if you compare the price you also have to look at the quantity of OPC in each capsule. There is major difference between 50 mg and 100 mg in the capsule. But these are not the only arguments – think about the filling station!

Is our special filling station bothered if "Ruedi Rüssel," "Shell" or "Aral" opens a filling station on the same street? No way! We do not have any competition, and our singularly outstanding feature is that we can achieve our goals by consuming a product that we need anyway (I hope that you can convince yourself of that, at least!)! It is very important that you use the last argument **only** together with the others.

And finally. How long will your new partner take with his products before his own products are paid for? I hope that you will help him to achieve this goal quickly!

"I don't enjoy doing the things that I have to do in Network Marketing."

I once read an extremely informative article on this subject entitled "You don't have to enjoy it…."

When a successful Networker was given this argument he said to his interested party:

"Do you seriously want to tell me that you have enjoyed being startled from your sleep by the ringing of an alarm clock for the last 30 years? Did you enjoy racing off in the morning and then being stuck in a traffic jam and breathing in other people's exhaust fumes? Did you enjoy working with a bunch of negative people, weaving their webs of intrigue throughout the company? Did you enjoy being told when you were allowed to take your lunch break, take holiday, when you were allowed to be sick and how much money you were worth?"

He concluded by saying that there were a lot of things that he would not like doing in Network Marketing, but he would like it even less if he had to work for 30 or 40 years. *"And if I do have to do things that I don't enjoy, I would rather do it for four years than 40!"*

And at the end he came out with a sentence that really touched me: *"But don't be surprised if you do not just enjoy doing the things that you have to do, but end up loving them."*

I have already told this story many times. That is what it is all about – do I want to work for four years and achieve my goal, or wait for 40 years! Do you prefer the 4-year or the 40-year model?

"Market Saturation"

It is almost impossible to saturate the market because we are working with consumable products. If you want to put it this way, many major companies have achieved "market saturation" – they are not taking on any more employees and are just concentrating on selling their products! After five years, my team is now about 30,000 users strong. Throughout Europe. And I have only counted the ones who actively place orders. How many people live in your city? You see! THER IS PLENTY POTENTIAL!!!!!

"Where's the catch?"

People ask time and time again about the "catch." My standard reply to this kind of question is *"You have just found it!"*

To explain this I would like to tell you a story:

Imagine that you are in a room, and the speaker has a 100 Euro note in his hand and says: *"I am selling this note for 10 Euros. Who wants it?"* What would you do? Jump up and shout *"Me!"*? Or wait and see what the others do? Unfortunately we are taught in life that only "idiots" fall for something like that.... And therefore we expect a catch. Who can believe that someone would make such a fantastic offer? And if something looks as attractive as this, there **must** be a catch in it.

Sometimes I wish that there was a catch – at least a "tiny little one".... Perhaps it would then be easier, because people would discover it straight away and accept it. But what do we have instead? A possibility of achieving **EVERYTHING** that we want to. Without limits: visions, dreams, health and leisure. And all without risk. And what do you have to put into it? Depending on your goal, it is a matter of how much time you need to invest. And a few Euros for petrol and books, which are probably tax deductible....

You will discover the catch when you look in the mirror every morning.

Stable income from the bottom up

You heard about the bigger numbers a couple of chapters ago. For this reason I am now going to give you an insight into the **depth**.

A company whose marketing plan is exclusively based upon Referral Marketing can pay out more at lower levels than companies with a direct selling component.

Let me explain it to you: Let us assume that on average, 60 % of all money is spent on marketing. Here are two examples:

Company one gives 30 % discount on direct sales or sales and pays 30 % to lower levels for the organizational set-up. Let‘s say that this is distributed on six levels, each receiving 5 %. In reality this is very rare, most companies don´t pay so many levels. Company two, which does nothing but Referral Marketing, does not do any direct selling and can therefore pay out the entire 60 % to lower levels. Let us assume that this is distributed on 12 levels, each level receiving 5 %.

You will see that the amount paid out by the company is the same in both cases, i.e. 60 %.

It doesn't make any difference to the company. **But it does to us**.

95 % of my income is currently generated at level five and below – imagine that our marketing plan only paid out four or five levels deep because of the high proportion of direct selling, like many other plans?

A compensation plan that pays out at lower levels can be a real gold mine that will give you a stable income that will continue to grow over **the passage of time**!

You can even see from the compensation plan that you will meet friends in Referral Marketing who will help you to talk to their friends about this fantastic opportunity. Karl Pilsl said:

> **"We have to make people realize that we do not want anything FROM them, but want to do something together WITH them."**

To give you an idea of the size difference between levels five, six and further down, let us simply assume that everyone sponsors five people, and that this is done beyond the fourth level, which takes time!

Level	1	5	Annes
Level	2	25	Bernards
Level	3	125	Christines
Level	4	625	Davids
Level	5	3.125	Emilys
Level	6	15.625	Fredericks
Level	7	78.125	Gertrudes

I have only calculated up to level seven in the example. Since this is intended to be a practical book, I only want to work with figures that I have already achieved myself. In Summer 2004 I had about 30,000 people in my team. This is a figure that I am speaking of from experience. Everything beyond that is (still) theory – I will leave the rest to your imagination.

I know that it is something that is difficult to comprehend, but I am merely using this example to make you understand that the more levels at which a company can pay out, the more income you will receive.

Strangely, this tremendous advantage is not always evident to begin with. I too thought of a disadvantage when I was first introduced to this plan. With my previous company I was given an extremely large discount on my own products, and to begin with I was somewhat upset to discover that this was not the case here. But gradually I came to understand, and I came to appreciate the tremendous advantage.

Training Meetings

The majority of Network Marketing companies have so-called "Meetings," which are also referred to as "information meetings" or "sponsor meetings." These are meetings to which new partners can invite their friends and acquaintances to watch the presentation and obtain information about the company. These are also known as "recruitment meetings." This means that the speaker welcomes you with the words *"Good evening ladies and gentlemen, I would like to welcome you in the name of 'XY'. Today we are going to introduce you to a business opportunity that... etc."* Strachowitz spotted the disadvantage of this method:

"Unfortunately this method lulls some partners into a false sense of security, and frees them from the responsibility for their own business. Dependence on the organization's schedule also plays a part – if the interested party does not have time on the day of the business presentation, it may be a couple of weeks until the next opportunity arises, which is a built-in growth restrictor. The aim is for every partner to carry out conversations themselves."

My previous company also held meetings like these, at which I soon became a speaker. And that was often a challenge for me! Visitors were often invited under false pretences and were extremely suspicious of the business, particularly because of this fact. You cannot imagine how difficult it was to stand there and have to convince these visitors.

I saw immediately that I did not need meetings to explain this opportunity, whereby I did not have to sell any particular quantity of the product and did not have any starter fees to pay.

It was immediately clear to me that I could explain it to any of my friends in no more than an hour. And they to their friends. The epitome of duplication! I realized that I had a simple opportunity that anyone can show to anyone else and immediately decided that in future, I would only speak to people who **wanted** me to speak to them!

I found an interesting chapter on the subject of "Meetings" in Don Failla's book that convinced me immediately:

"(...) a typical information event looks like this: rows of chairs in a living room or hotel room, and a board or flip chart at the front. A speaker in a dark suit introduces companies, products and the marketing plan, of course. These events usually last for about one and a half hours. Of the 22 people that are present, 19 are already partners, and the other three are visitors. More were invited, but they didn't turn up. The speaker addresses the visitors. He or she is only speaking to 3 of a total of 22 people! Most of the advisors that are present are struggling to keep awake because they already know how the event goes and could recite it in their sleep. At some stage it becomes so boring that they don't want to come any more."

I almost killed myself laughing.... That's exactly how it was for me!

And of course, I didn't want to do it any longer. Apart from the disappointment that a new partner feels when their guests don't turn up, this method is difficult to duplicate.

The majority of people are more afraid of the idea of having to talk in front a large number of people than they are of dying.

And what will your guest think? *"Help, in future I will have to sit in a meeting every Tuesday evening. I really don't have the time!"*

As I have already said, two questions are stamped on the forehead of every potential new partner. The first is: *"What's in it for me?"* and the second question is: *"Can I do it too?"* or: *"Do I have the time?"* My experience has been that these meetings are not particularly interesting to our new partners.

Just think about it: **Time is precious!**

For this reason we hold training meetings. The main difference here is that we are talking to the 19 partners! Guests can be brought along, of course! But with totally different prerequisites. Ideally, the guest will already have read some independent literature and is aware that he is a "guest" and can simply look in without any obligation. How much more credible do you think that this is than explaining to your active partners that they have to help the new guests in order to be successful themselves? They feel under less pressure – I know from some guests that they did not feel particularly comfortable at a recruitment meeting with the above-described format. They felt that most people were already "active" and thought that the entire event was staged for their benefit.

The pre-information can be practised at the training meeting, as can the starter training, and the questions that have arisen since the previous meeting can be replied to, of course. An exchange of experience can take place and (what is extremely important) one or more partners can therefore be speakers for an entire town, without giving the impression that only those who can speak in front of a groups of people can be successful.

Our open house meetings are rather there for an exchange of ideas and experiences. If the group is small, the people who are present can **briefly** introduce themselves and perhaps also explain what they expect to gain from this evening. I see the role of the speaker as that of a "presenter." The people who are present can be involved in the outstanding questions and therefore actively integrated. The learning effect is much greater than if a monologue is held. It's also more fun! The meetings ultimately have a certain "hospital function." Everything does not always run smoothly, and it is therefore extremely important for us to meet other people and see that they also receive "**NO**" replies, and that this is perfectly normal.

This is the path that we have chosen at *GabisteinerTEAM* and it clearly depends on our intended target group. We wanted to find a solution for the majority of people who are not salespeople, and usually do not want to give talks, or cannot. For Referral Marketing from person to person. Of course, there are teams that mainly consist of therapeutists, non-medical practitioners and similar professions and therefore operate in a completely different way than giving talks. This is perfectly OK, but logically speaking it is unsuitable for our target group.

The most important thing of all is enthusiasm, which has an infectious effect. And identification. Everyone can find someone at a training meeting who is of the same age, comes from the same town or the same country, or has the same profession. This means that everyone finds someone to identify with. This is why "Promotion," i.e. "moving a person from point A to point B" is so important. Particularly at our major events, which take place about four times a year. We now know from experience that it saves us at least a year of convincing work if our new partner has seen the "bigger picture" at an event. Once they have met the founders of our company. This is why we promote these occasions with all our might!

Training meetings and GabisteinerTEAM

When I started out with Lissy and Isolde, our company had little involvement in Germany. Time and time again I hear arguments like *"It's OK for you, you were there right from the start!"* That is a misconception!

Isolde sponsored Susanne and Susanne sponsored someone from Münster in Westfalia. This means that Isolde and I went to Münster and did some training there. On the way back we made a stop in Siegen. And the same then occurred in Bremen, Berlin, Munich, Salzburg etc., etc., etc.

These training meetings now take place all over Germany. And in Switzerland and Majorca. We do not **have to** travel as much, which is a giant leap forward and provides an unbelievable amount of growth potential! Everyone has the opportunity to look in the online calendar in our member's area and send a friend or acquaintance who lives somewhat further away to a training meeting. Of course this is only possible because all the training meetings adhere to the same system.

At this point I would like to re-emphasize that the partner we send to a training meeting has been sponsored and had a starter training. It goes without saying that the trainer does **not** have to explain the basics to our partner. It also needs to be completely clear: I prefer to support a really serious partner myself: "Learning by doing".

GabisteinerTEAM is an amalgamation of the partners in my downline, as well as sidelines, who respect each other through loyalty and on the basis of common ethical principles, and help each other to operate much more efficiently and globally because of this cooperation. I would now like to briefly explain the guidelines upon which *GabisteinerTEAM* is based. The basis of *GabisteinerTEAM* is a system that we have worked out together, adhere to and do not change independently. It has been tried and tested in practice and provides a uniform way of working and of training our partners. The main constituents are the global system, the catalogue information, the starter training and independent tools. These are documented in the member area and can be read and printed out by anyone. We do not think that we need to use inserts which promise millions, nor distribute flyers and business cards at random ("cold contacts") to people. We prefer to operate in the "warm market," or first build up a relationship. We use our personal story, and in the ideal case we only pass on information if the other person is interested. In this way we avoid rejection and do not "burn any bridges." The ethical rules also mean that we do not steal partners from other lines or criticise other partners, companies or products.

Filtering and Sorting

I share the opinion of many successful managers in the industry concerning another issue:

> **"The key to immense growth lies in finding leaders and training them!"**

This means that our greatest challenge is to recognise this. Finding out at an early stage **who** is really interested in becoming successful. Time is our most precious asset, and I think that you now understand what can happen if we sponsor an active partner and help them to build up their own group by providing active assistance in doing the same training with their group.

Filtering and sorting starts at an extremely early stage. For example, I am already filtering out who is interested and who is not by telling my story. If I already have a few partners in my team who are in the "Not-Yet-Box," when I have time again I can filter them again by asking *"I currently have some time available to help you and make you successful. Which one of you would like to be that person?"*

If we **really** do it like this we are in an absolute "offer position," in which case our quota will increase in leaps and bounds....

For this reason it is also important not to waste time on **asking** people to get into our business. Reluctant people will not be good partners – even if we can convince them to get started. The people who we are looking for are the busy bees, the people who start work immediately and not *"after the summer holidays..."* like some people.

I like to compare it with falling in love. Could you ever imaging saying to your lover *"It's no good in summer – let's wait until it gets cooler...?"*

Take advantage of the initial enthusiasm! Speed is one of the greatest secrets in our business. But don't plan it with too much logic, otherwise you will lose the power. And don't think you have to read all the books before you get started. Filtering and sorting means that we support those who really want to and do not waste our time on people who are not really interested.

You can also use tools for filtering. As I mentioned previously, tools are books, videos, seminars and unbiased newspaper clippings etc. I hand out a suitable tool for the type of person and the target group in- volved. By means of pre-filtering with a tool I can find out whether or not there is any real interest, and not necessarily after a three-hour discussion!

Because you must always consider: **what kind of example am I setting?** You are only credible and authentic if you practise what you preach. For example: *"You can also set up the business on the side."*

One thing we do know: we are rarely approached by our interested parties. You know how it is – how many notes do you have on your notice board concerning things that you really wanted to do or books that you wanted to buy....

As soon as I hand over the tool, I suggest to my partners that we talk about it: *"What do you think – when can you read the book?"* or *"When can we meet up again to talk about it?"*

Logically, the quota is higher if I arouse curiosity beforehand or have appealed to my partner's **WHY**.

And if someone cannot decide to meet us? Good, then they have made their decision and filtered themselves out! That is perfectly OK. We know that we need our quota and do not lose any of our enthusiasm because of it.

In the book entitled "30 Most Dangerous Documents" by Ledoux, I found the following passage concerning **mistakes** that are made by almost everyone:

"Someone gives up and you are sad. But remember: Referral Marketing is a numbers game. A marathon. A stamina competition. 50 % give up in the first year. 40 % love the products. 10 % do the business. 2 % do 90 % of the work.

Keep an eye out for winners and be one yourself, and the winners of the future will also find you."

I find that magnificent! That is precisely what it is about: finding this 10 %. No more, no less. I think that looking at it in this way helps to combat disappointment. Please never complain that "he" or "she" is not DOING anything. That is up to them! This is how we offer our business (or at least I hope so…).

We say: "You can do, but you don't have to!"

In the early days I had a teacher in my team who asked me whether she and her husband could talk to me in private. I thought that they wanted me to give them some training. I was therefore somewhat surprised when her husband sat at the table, crossed his arms and said:

"So, Frau Steiner, try to convince me – and I can tell you now, it won't be easy…." I couldn't believe what I was hearing, and said: *"In that case I'm not the right person to speak to. I do not want to convince you, nor will I attempt to. That would be a JOB, and I don't want a job any more. I would be pleased to recommend some literature that you can read in your own time, and I can reply to any questions that arise. But you have to convince yourself. And once you are convinced, then you can convince me that you are a serious partner and that you are worth investing my time in!"* I didn't put it quite as blatantly as that, of course – but they understood what I meant.

In her book on this subject, Paula Pritchard came up with a wonderful example: *"I imagine that I have discovered a gold mine and only have five spades. The question is 'Who should I give them to?'"*

I like this idea, and it hits the nail on the head – but only if we are really prepared to support our partners! But how do I recognise an active partner?

In the chapter entitled "How to recognise a gold mine," Don Failla writes:

- "He is prepared to learn, has 100 questions and keeps calling you in the starting phase" (that's right – I become anxious if a new starter doesn't have any questions)
- He asks for assistance and wants you to get to know his interests, which is an extremely important point!

- He is totally enthusiastic, reads books and learns everything about Network Marketing in order to see that it works.
- He is prepared to commit himself, and buys and uses the products.
- He has goals.
- He has a list of names and uses it.
- He is fun to be with.
- He is positive. Everyone likes to be with positive people.

To this list I would add:

- He uses the tools to convince himself and then uses them with others.
- He invests time and travels to get-togethers and meetings come rain or shine.
- He is prepared to take responsibility.
- He works through his self-imposed daily work quota in a disciplined way.
- He sees obstacles as a challenge.

You cannot make managers – you have to find them.

A sponsor's duty

One of the most important principles in Referral Marketing is:

THEY DON'T DO what YOU SAY – they do what you DO!

It is therefore important to be a role model for your team. This means that whatever I tell my Downline to do, I must also be prepared to do myself. Another important point is finding out which partners are serious and which are not. A good sponsor takes their serious partner by the hand and helps them to build up a team.

"Learning by doing" is needed. **Personally, I respond to questions as soon as they arise**. Believe me, I could feed my new partners with theory all week long – I have enough material.... But I am certain of one thing: **the start must be simple!** As a good sponsor I keep in touch with my new partners and keep a note of **when** the next **session** is due.

Ultimately, one thing is certain: you have to learn and develop if you want to reach the top! But this occurs automatically and will not take place within the first few months. I often say to my partners: *"When you first start working for a company, you do not have to know what you will have to do as the general manager!"* That is fairly logical to me, somehow. Lissy also has the saying:

"A good sponsor is not always an easy sponsor to get on with!"

A good sponsor must also be prepared to deal with unpleasant matters that are obviously preventing our partners from being successful. The "MLM Coach" formulated the matter quite nicely in a circular:

"If your partner vehemently refuses to introduce you to their interested parties, it may have something to do with your appearance.... To be honest: I would think twice about introducing my interested parties to a sponsor who is 'grumpy', has greasy hair or whose living room is hardly visible through clouds of smoke."

At this point I would like to pass on a snippet of information to my fellow smokers: smoking at meetings or events can be extremely costly to us. After all, we are dealing with health products. I have received a great deal of feedback from guests who were at smoke-filled meetings and decided that it wasn't for them because of this alone, *"because the people have no credibility."* And you will probably never discover the reason.... As we have already said:

You cannot measure lost opportunities!

To the question *"Who should I support?"* there is an interesting aspect on a CD by Jim Rohn that I would like to recommend. He says: *"If you take one step, I will take two! If you take two steps, I will take three for you!"* That is extremely good, and I think that this has been so succinctly put that no further explanation is required.

"You can help 1000 people, but you can't carry three on your back!"

How true – and also frequently a problem! I know that many partners, particularly new ones, try to help people who "need it." It doesn't make sense to look for people who need help. It is better to look for people who **WANT** it.

Sponsoring and caring

You now know that because of our compensation plan, you do not continuously have to sponsor new partners. It a case of: sponsor and promote, sponsor and promote, sponsor and promote.... It is totally counter-productive to make ten people enthusiastic about the business at the same time and then let them "muddle along" without any support. Personally I think that it is better to devote 50 % of my energy to two lines that both have a good chance of being successful. If I devote 10 % of my time to ten lines, it may be that **NONE** of the 10 receive enough help and support to becomes independently active.

"It is every sponsor's responsibility to pass on all of their 'business knowledge' to the advisors that they have sponsored: Ordering products, the initial steps, how to build up and train their own networks etc."

Don Failla thinks that it is a major waste of time to keep looking for people who we can sponsor instead of helping a friend to speak to his own people. Of course, we have to spend 100 % of our time on sponsoring new partners when we first start out. Later you will spend less time sponsoring new partners, since we are not selling. The rule of thumb with us is: spend 95 % of your time with the serious 5 advisors, 2.5 % looking after other advisors and 2.5 % looking for new partners. Don Failla writes:

"If you train your advisors accordingly, you will never have to look for interested parties. You will get to know enough people to whom you can introduce your business whilst you are looking after your new partners, meeting them etc."

The more intensively we can work together with our partners, the more successful we will be. And precisely because of this an intelligent Networker builds up his group in his immediate surroundings.

We help out new partners to reach the third or fourth level – we spend most of the time that we devote to Network Marketing on this activity,

This is the key to success.

Please read this paragraph again! It makes a major motivational difference if I help my new partner to sponsor downwards rather than sideways, which I would like to show on the following example:

Let us assume that I have sponsored Anne, and I carry out four conversations and help her to find four new partners (Bernards).

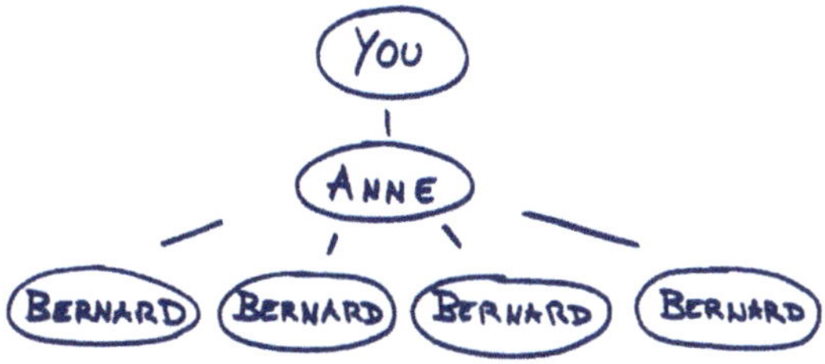

There are now four partners in Anne's team. Now, how many of them are motivated? If I assume that the best motivation of all is to have a partner of your own in the team, Anne is the only one who is motivated. Now I do the same amount of work and carry out four conversations again. Only this time I help Anne to talk to Bernard, then Bernard to Christine, Christine to David, and then help David to speak to Emily. Now let us simply assume that everyone starts. Anne now has four people in her team. But how many are motivated **now**? **Four!** 400 % more motivation with the same effort! That is the biggest secret in our business.

Don Failla has the following to say about this:

> **"You can motivate people faster and more effectively by putting a candle under their sitting place than by putting a blow-torch to their thinking place."**

Nothing motivates people more than having someone who starts "below them." Working the depth is priceless. With the same amount of effort I now have four times as many motivated people. The quicker you bring your own partners to the third or fourth level, the more you will fan the flame of enthusiasm, because our partners can see: *"It works! I can do it too! I enjoy my work! I am earning money!"*

And if this team is deep and wide enough and can also operate independently, before you start concentrating on sponsoring new starters again you can ask the people who are in the "Not-Yet-Box:" *"Hi Alfred, I have just helped a team to achieve success – they are now independent and I have time to support a new starter. Are you interested?"*

It is also a challenge for me to find potential leaders who are prepared to accept responsibility. A major advantage of our compensation plan is that the next star does not have to come from the top three levels. It can come from a partner at level ten or even further down. That is the reason why it is extremely important to spot active partners further down and contact them or "collect pearls from the deep."

I am pleased when anyone comes to me and asks for help. Of course, I can't do it all myself. But I will ensure that everyone is put in touch with a successful team. With our system and event calendar, it is no problem whatsoever!

Networkers - Top Networkers

The major advantage in Referral Marketing is that anyone can build up a second income at their own speed without having to leave the security of their existing job. The amount of time that they invest will determine when they reach their goal.

This book is intended for new partners. For this reason it is important to mention that the normal way is to invest between four and ten hours per week to generate a secondary income and later (four to six years) turn it into a main income. The monthly income of 30,000, 60,000 Euros or more that top Networkers earn is impossible if you only spend five to ten hours a week on it! If you have a bigger target or simply want to move upwards more quickly, ask your sponsor about power groups.

Closing Remarks

The submission date for my book is getting closer – it is unbelievable how many people have already asked about it and can't wait for it to finally be published....

This makes me extremely happy. I know the potential market for this book. For this reason I have been thinking for days about whether I have forgotten anything or whether there is something that still remains to be said. If anything is missing it will be in my book for advanced networkers that is already in the pipeline.

I hope that I have been able to give you a few arguments as to why you should start, and that I didn't give you too much information at once. If so, forget everything that you don't need at this moment and do it in a way that you feel comfortable with!

Perhaps you will read the book again in a few weeks or months. I will guarantee that it will then seem like a totally different book! I am not the only one who has experienced this. Because you only read and take in the parts that apply to you **at the time**.

Just ensure that you always have a good gut feeling. If that is not the case, consider what you are currently doing. Perhaps you are trying to persuade someone? Or you are involved in discussions with someone who has a negative attitude towards the industry? I always say: *"If you do not have a good gut feeling you are doing something wrong."*

But that is not a problem either. The boss of IBM was once asked how to get to the top in his company. He said *"Double your error quota!"*

> **"Sometimes you don't spot the paths until you start walking."**

Consider this: you step onto new ground and need time to become "experienced." It was no different for me and my team. We made decisions, implemented them and watched what happened. What was good? What had to be corrected? There will never be a solution that is perfect for everybody.

Nor is there a **single, perfect sentence** or a **perfect way of talking to people**. This is something else that we have to live with. What I can promise you is that **you** will improve with experience!

At this point I would like to give special thanks to everyone who has accompanied me over the years and provided support and assistance during the writing of this book. We never stand still and we never will. A great deal of flexibility is required! This is also the reason why I have not made any detailed recommendations in the book concerning the training system. The latest state of the system and the recommended tools can be found at www.mitgliederbereich.com.

What specific suggestion can I give to you when you start?

Consider the **WHY**. Then make a decision. Tell your sponsor about this and arrange a date for the starter training. Invite your best friend to it. Order the set of books that suits your goal. Make a note in your calendar of the dates of the events that you are going to visit, and the ones at which you will be accompanied by your sponsor. Take your new business seriously. Even if you enjoy it. And then commit yourself to sticking with it for a year, regardless of what happens. It was none other then Goethe who said the following on this subject:

Unless you are committed you hesitate, and run the risk of taking a step backwards and always being ineffective. There is an elementary truth that applies to all initiatives and creations, and failure to be aware of it can destroy a countless number of ideas and magnificent plans. In the moment that you irreversibly commit yourself, providence will also become evident. All possible things occur to help you that would otherwise never have happened. An entire flood of occurrences flows out of the decision, invokes all possible unforeseen occurrences, coincidences and material help to your own benefit, of which nobody could have dreamed would happen to them. You are capable of doing anything that you can imagine.

Start now, straight away!

There's a lot in it – it is confirmed to me time and time again – and after all, Goethe should know....

My final tip:

Recognise the value of people!

If you have someone who seriously wants to do it, give them all of your support. Be aware of what you might mean to this person and what this person might mean to you! I would like to leave you with a quote on this subject that is extremely appropriate to Referral Marketing:

"Anyone can count the number of pips in an apple, but not the number of apples in a pip!"

I wish you all the best in your life and I hope that your decision will be in favour of joining this marvellous industry. I would be extremely proud if I felt that I had contributed in any small way, and I am looking forward to the next event at which we might meet!

Kind regards,

Gabi Steiner

February 2013

After 16 years we are back in South Africa. I have work with me – the English book has to be edited and re-printed and in a hurry. And lo and behold! All of a sudden the Universe has given me the necessary time for it...

Due to the snow in Munich we were delayed and missed our connecting flight to Johannesburg, South Africa. This meant one day in Frankfurt and plenty of time.

As I read the English book, I realise that so many things that have changed in the last 14 years. Incredible...

After reading a few chapters, I am flabbergasted: I can read the book WITHOUT a dictionary! This was NEVER possible before – I always had to have a dictionary beside me and, because of that, I usually lost interest after 10 pages...with the result that I never actually read it myself.

And now I read the English book like I read the GERMAN one! Wow, I was not aware HOW much my English had improved since I helped Rachel to build up her team...

However, because of this, I also found mistakes in the translation, some of which are distorting the MEANING! Not to mention all the changes that have taken place since this book was first published!!!

In the plane I thought about it a great deal and decided to just eliminate the worst of the mistakes and add a new chapter at the end incorporating all the changes that have taken place in the past few years. And that is a lot!!

What surprised me and made me happy is: THE CONTENT – the basics and, therefore, the system are still the same and I am delighted about the fact that in Germany everybody is just as enthusiastic as in the beginning.

At this moment the whole of Germany and, of course Switzerland, are working to the motto: BACK to basics!

Well, what has changed…

In the meantime, we spend a lot of time in Majorca. Lissy and Werner, as well as Isolde and Ottmar, have discovered the advantages of the island and spend as much time as possible here. We hold regular meetings and training sessions which our partners from Germany appreciate immensely in order to recharge their batteries, not only with inspiration, knowledge and experience, but also with a little bit of sun.

This is possible, last but not least, thanks to my brother Andy. Behind the scenes, he runs the entire show which means he is the backbone of the company - my right AND my left hand. I wouldn´t know what to do without him…

Along the way, he has built up his own team and achieved an income he could only dream about before…HIS personal development has been phenomenal. In January 2013, he and Uta hosted the Kick-Off with 1,200 people at the "Schwabenlandhalle". For someone who was once a shy person, this is a proud achievement! Our mother was in the audience. You can find her latest photo at the age of 77 on my website.

Of course, not everything is rosy – misfortune has hit Erika, our housekeeper with 5 children particularly hard. A few years ago her son had an accident and lies in a coma.

This is very hard and I think everybody can imagine that there is no solace. Nevertheless, I am very thankful that she can dedicate her time to the care of her son and does not have to be concerned about her income. Her team is independent and I know how much she is looking forward to the cruise every year to escape her daily challenges for a few days.

My son has almost finished studying and at the moment works at a sports studio, something he loves doing. I am VERY proud of him !! AND he has become a vitamin freak.

On page 46 you will find Georgia´s story, a mother of 6 children. A few months ago, I met her again for the first time in years. And guess what! She wants to start working actively, now that her cheque is about 2,000 Euro. How did this happen?

Years ago she sponsored her midwife and Cordula has build up a big team, together with her husband Werner, and today she, like so many other leaders, is my friend. WHAT A BUSINESS !!!

We also have changed the name of the team. We are just the *GabisteinerTEAM.*

Well, and why has my English improved?

When we bought our house in Majorca in 2005, I did not want to work as much any more. I began to play tennis, which has become my great passion.

In 2010 I saw that Rachel, an English lady with whom I never played because she is just so much better than me, was losing a lot of weight. I was told that she was going through a divorce and desperately looking for a job…

Mmh…what should I do? Especially since I knew I could help her… But…!

What would happen if I started again with her and she did not succeed?? I was concerned that I would lose face and have to change clubs J. And not only that! I was quite aware of the fact that I had to hold meetings for her in English !!! Last but not least, I was totally aware that: IF I worked actively in Majorca, we would go international in a very short time!!! Majorca is a kind of microcosm of society where ALL nationalities come together!

It was very clear to me that this meant work. And not just a little... Anyway, I spoke with her and I gave her my book. She took advantage of the opportunity. Well, now you know why all of a sudden I can read my own book in English with no problem… Of course, in the past few years I have held a lot of meetings and training sessions in English…

A small change that my company made to a product came to my aid – and this was such a hit that today it has become a strategy, which is NOT YET described in my book:

In 2010, a skin care product was enriched with hyaluronic acid !!!

This came just at the right time!!! As you well know, our strategy is to work with INDEPENDENT tools ! All well and good for Germany and Switzerland! But: neither in ENGLISH nor in SPANISH did I have any independent tools!!!

And suddenly there was the hyaluronic acid in the Skin Formula! All of a sudden, I had the brainwave to hold a "before and after" photo competition - with fantastic prizes, of course! All the participants had to send in a photo and then an update every 2 months. This became a big hit! After 2 months we had 600 participants and photos that practically blew me away! By the way, the photos of the winners can be found on my website.

This was a decisive breakthrough! I realised immediately what an opportunity we had missed all those years before!

It became very clear that every partner HAD to have a good BEFORE photo without make up. What a great achievement when someone could demonstrate visible improvement from using the products – and, believe me, they WILL experience it!

Today this has become an important element in our system.

And what is so valuable: photos do not need to be translated !!!! They speak every language ! And that is necessary – here is an email from Rachel I just received while in South Africa:

Hi Gabi,

I forgot to tell you: last Thursday we had a meeting at Deborah´s with her downline.

Rachel (English) well actually Yorkshire

Deborah (Bermuda)

Camilla (English)

Daphne (German/French)

Susi (Italian)

Miki (Danish)

Eva (Eastern European)

Anna (Spanish from Sevilla)

Now that´s international. All in one line in one apartment !!!

Oh, and 3 dogs, one white, one brown, one black!

Rachel x

What a story !!!

Last year Andreas Müller, a leader in my team, spoke about an old friend of his, a totally crazy guy from Surinam who lived in Holland, whom he had finally sponsored after 8 years. He was totally enthusiastic about him and, when I met him and his team last September in Majorca, I knew what Andreas meant! This guy was the epitome of power and passion – all of you who have listened to the calls where I have invited him to speak, know what I mean.

So, now I have agreed to a first training in Den Haag in March – and that is why I need the English book!!!

Not only Holland will be strong – Rachel will have the first big meeting in England in April together with Ulrike. All the other countries will eventually follow. It is just a matter of time.

In the meantime, our company provided us with a website, free of charge, which everybody can personalise. At the click of a button, it can be changed into many languages – even into Japanese! What an opportunity! Although I still believe that there is NO substitution for one-on-one contact with people!

Another important thing happened in 2011… Action 32 was born. This came about because of Elke. One evening we sat together and I asked her: "why don´t you do what you know?" She answered "Actually I don´t really know what I have to do!"

That´s when we SAT DOWN to actually "DO". And since then it has worked really well. Her personal development in such a short time is tremendous. Especially NOW she is a great help for me in applying this knowledge in real life.

Elke is also a good example for how much tolerance is possible in the network. She is the new wife of my son´s father and, precisely for that reason, it is important that she has confidence in me and allows me to guide her. She is one of my numerous partners who are unstoppable on their way up.

The number 32 in Action 32 stands for ENOUGH! A deck of cards has 32 cards with 4 aces… If we just turn over 3 cards, the possibility of finding an ace is relatively small. Even after turning over 15 cards, it is still possible that you will not find an ace. In a worst case scenario, it could be the last 29 to 32…

And this shall be our basic rule. No matter what we do, we do it 32 times… and this at

least with a "5-10" Well, I don't want to reveal what all of that is. It is better that you ask the person who gave you this book.

It was a light-bulb moment that I have described exactly in this book, but it took soooo long to realise it:

DON'T JUST TRAIN your partners – ACTUALLY DO IT with them!

I described it in this book!!!! RECOGNISED and IMPLEMENTED IN A SYSTEM in January 2013.

Since then Germany is on fire…

And I had to go to South Africa J

With warm regards, wishing you great success in everything you are doing!

Gabi